The Bactericidal and Oligodynamic Action of Silver and Copper in Hygiene, Medicine and Water Treatment

Co-authors of the German Edition:

Günther Beeh (Karlsbad)
Hartmuth Bohmeyer (Alveslohe)
Hasso Kaiser (Schwäbisch Gmünd)
Alfons Knödler (Schwäbisch Gmünd)
Frank-Detlev Koppitz (Gomaringen)
Frank Krüger (Bruchköbel)
Christoph Raub (Schwäbisch-Gmünd)
Volkmar Rössiger (Freiberg)
Kurt Schindler (CH-Feldbrunnen)
Andreas Zielonka (Schwäbisch Gmünd)

Medical Consultant – Rolf Breitstadt (Bad Soden)
Water Treatment Consultant – Siegfried Hauswirth (Grevenbroich)

The Bactericidal and Oligodynamic Action of Silver and Copper in Hygiene, Medicine and Water Treatment

Dr Uwe Landau and Co-Authors
English Language Edition Translated & Extended
by Dr Anselm Kuhn

Finishing Publications Ltd, Stevenage, UK
www.finishingpublications.com

German Edition ISBN 3-89995-284-7. Isensee Verlag, 2006

This Edition: ISBN 978-0-904477-30-6

A catalogue record for this book is available from the British Library

Typeset in 10.5/13pt Sabon by Integra Software Services Pvt. Ltd, Pondicherry, India.

To the cousin I never met, Professor Reinhard Kuhn (1931 to 1981), Brown University, USA, who died of Legionnaire's Disease while on vacation in Italy

Contents

Foreword to the German Edition

The Noble Metals Committee of the DGO, the German Association for Metal Finishing and Surface Technology set itself the task of producing a textbook covering the basics of noble metal deposition (1a). Having done so, they see their brief as dissemination of information on established and new coating processes and related topics.

The present volume describes the oligodynamic effect of silver as a bactericide, both from a historical aspect and in terms of more recent advances in the understanding of its workings, as well as the practical implementations and applications of the concept in combating microorganisms in water treatment, medical and more general contexts such as catering and food preparation. In these and other applications, the metal can be used in continuous form or in particulate form, loaded into paints, polymers and other substrates. For this reason alone, it is right that surface finishers should concern themselves with the bioactivity of the metal. Equally, the subject should be of interest to metallurgists and materials scientists.

Of the metals which concern our Committee, silver is foremost in terms of its bioactivity, in which there is a growing interest. Our aim, with this volume, is not only to promulgate a wider knowledge of, and interest in, the oligodynamic properties of silver, but also to provide the information to stimulate its use by manufacturers and industry in novel forms or applications. It is our aim to encourage coaters, whether using electrodeposition or gas-phase technologies (such as PVD) to add products making use of this phenomenon to their product range.

Grateful thanks are offered to all the authors for their input, made despite their other professional commitments. Their work included researching the literature and critically evaluating it, before writing their respective chapters. Gratitude is also expressed to Dr Rolf Breitstadt, who proof-read the text as a qualified medical professional, making valuable corrections and additions in terms of the medical aspects and for improved clarity. Special thanks are also due to Dipl. Eng. Siegfried

Hauswirth, who so readily agreed to check those parts of the book related to water treatment. Thanks to his input, corrections and additions to the text were made possible, as well as assuring the correctness of the complex issues related to water quality legislation and practice.

Last but not least, special thanks are due to my assistants, Frau Schuster and Frau Valentin-Biermann, who assembled the final manuscript and who incorporated the numerous additions to the text with such patience.

Berlin, July 2005 Uwe Landau

Foreword to the English Edition

Changes and additions have been made, reflecting the rather different demands of English-speaking readers from their German counterparts. In the UK and the USA, as in Germany and Japan, there has been a surge of interest in the oligodynamic effects of silver and copper, prompted by the proliferation of often-fatal hospital-acquired infections caused by antibiotic-resistant bacteria. In the UK and USA, the emphasis is arguably tilted more to the study of copper and its alloys, than silver. Though the German Edition does also treat the former metal, some additional material relating to copper has been included in this Edition.

The intended readership of both German and English Editions is mainly those who wish to learn about the subject with a harnessing of that knowledge in mind. To that extent, it is clear they would wish to be as well-informed on the oligodynamism of copper, as of silver. The applications and scientific knowledge relating to these two metals are so intertwined that to omit a discussion of the one metal would be devoid of logic.

ORGANISATION OF THE ENGLISH LANGUAGE VERSION

This English language version includes a CD-ROM which has several components. Firstly, a fully searchable abstract database with over 500 patent and publication abstracts. Instructions for use of this will be found at the end of the book. Secondly, and thanks to those authors who have made their work available under "Open Access", one can access and reproduce their writings in full. Some such key papers are included on the CD-ROM. Readers are reminded of the conditions allowing this, which is that while these may be further distributed without permission or payment, the original authorship and source must, without exception, be acknowledged.

References: These are cited in the conventional way. The CD-ROM database should be used as an additional source of information. Separate instructions for this are provided with the disk.

Acknowledgements

Thanks are due to those who assisted in production of the English-language edition, including John Hayes of TarnPure Ltd, Professor Bill Keevil of Southampton University and Dr Peter Neufeld, all of whom provided valuable information and insightful comments. The Copper Development Association (CDA) and Silver Institute in New York have both been active in supporting and promoting, in various ways, the application of their respective metals to fighting pathogens. Both have web sites with much interesting and valuable information.

Stevenage 2007

PART 1

THE HISTORY AND SCIENCE OF OLIGODYNAMICS

1

The History and Science of Oligodynamics

1.1 INTRODUCTION – AN OUTSIDER'S VIEW

"Oligodynamics" . . . what an odd word . . . what does it mean ? What is this all about ? What sort of people are involved ? This book is an attempt to answer all these questions, and many more besides. The beneficent effects of silver and copper, known from the most ancient times, are recorded here as is the pivotal work of von Nägeli, arguably the man who took some two thousand years of folk wisdom and placed it onto a scientific basis.

The outsider soon recognises the many players in the field. At the centre, the mainstream medical profession, anxious as ever to combat and prevent infectious diseases or localised sources of infection. Beyond them, others, some medically-qualified, or with a scientific or technological background, investigating and advocating the use of silver- and copper-based materials for maintaining public health, as in water treatment and preservation of other beverages, control of bacterial infections in a wide range of institutions or hospitals or food-processing plants. All such activities are fully accepted and endorsed by "the establishment" world-wide. The same cannot be said for those, some medically-qualified, others not, who advocate silver-based therapies for a range of oncological conditions, viral infections and other conditions, which might not be endorsed by most of their professional colleagues. At the outer fringe, one finds businesses, some operating as "Foundations", whose aim is to sell silver-based preparations to valetudinarians, hypochondriacs and self-medicators. Some such Foundations may well be totally *bona-fide*. Readers can draw their own conclusions, for example on

IMREF, the Immunological Research Foundation Inc (www.imref.org) whose extensive website is worth a visit for its very catholic bibliographic list of references. The CD-ROM database includes internet information relating to some of these, often extensive websites. This should not be taken as endorsement (or condemnation) of their activities. Readers must form their own conclusions and make their own classifications as between mainstream orthodoxy, speculative science and outright "fringe medicine". Many years ago, a young scientist, lodging with a German family in Munich, was amused at the custom of his host who, every Sunday, would remove his golden wedding ring, boil it in water and then drink this potion. Who can tell where this custom originated, how widespread it is and whether it has any medical basis ? The assiduous reader might also encounter the world of homeopathy which, at least superficially, might be thought to incorporate the concepts of oligodynamics. No more will be said of this.

Other important players in this field are those involved in water engineering, whether for urban water supplies, in hospitals and other large public buildings and those involved in domestic plumbing. Of their invaluable contribution to public health, there can be no doubt, as numerous scientific papers testify. Last, and most certainly not least, are two umbrella organisations, the Copper Development Association which is active in many parts of the world, notably the USA, the UK and Europe, and the Silver Institute in New York. Both have websites www.cda.org and www.cda.org.uk especially www.cda.org.uk/antimicrobial/ with valuable technical information, as well as links to some scientific studies. Granted, they exist to promote their respective metals, but even so, the reader is strongly advised to visit their websites at regular intervals.

What use is all this knowledge if it cannot be harnessed ? The second part of this book focuses on the more applied aspects of the subject, both specific applications (mainly in the field of medicine) and in the development of materials – alloys, fabrics, ceramics which incorporate oligodynamic metals. But to capture a fuller extent of such applications, the CD-ROM database should be consulted. Here we find patent abstracts for medicaments, catheters, wound-dressings and other medical applications. We find domestic applications ranging from water purification, combating legionnaire's disease in domestic hot water systems or bathroom shower hoses. We find patents relating to air-fresheners and de-odorizing, pet hygiene, laundry and washing machines, bedding, clothing and shoes. Only browsing can really do this collection justice. Von Nägeli's work was published in 1893 and the CD-ROM database makes clear that, already in the 1920's, there was a stream of patents and scientific studies built on his work.

1.2 "OF THE MAKING OF BOOKS. . . ."

"Of the making of many books, there is no end . . . " (Ecclesiastes) goes the saying. And it remains true today. Yet in the first decade of this new millennium, change is very much in the air. Thanks to various technological developments, e-books, after several decades in gestation, seem set to become a reality. Equally important, is the trend to "Open Access" where complete scientific papers are available on the internet, free of charge for all to read. In other cases, extended abstracts are freely available on the internet, with the full text available against payment. The liberal use of URL's (internet addresses) in the text of the book and in the CD-ROM database, is intended to encourage and guide the reader to the huge resource of material "out there on the internet".

1.3 MEETING NEW CHALLENGES

Mankind has always faced fresh challenges and never more so than today. Climate change poses a threat that some distinguished scientists fear may lead to the end of civilisation as we know it. Uncurbed population growth, coupled with falling crop yields in some regions, is a looming scenario to which no politician has offered an answer. The end is in sight, of a unique era of low-cost, easily extractable and useable hydrocarbon fuels. The dire forecasts, subsequently much ridiculed, from the Club of Rome, that we would exhaust our supplies of naturally-occurring resources, now begin to be credible. Around the middle of the 20th C, Man was confident that the battle against disease was slowly but steadily being won. The problem was simply to address, slowly but surely, the conditions that existed at that time. The AIDS/HIV epidemic was undreamed of. So too were the lethal forms of influenza such as HN51 which threaten a global pandemic unknown since the Spanish 'flu epidemic of 1917. Not only were such diseases, new or recurrent, unimagined but so too was the expansion of low-cost global air travel. Epidemics whose progress could once be tracked in months as they moved across continents, can now spread within days. Likewise, drug-resistant TB or bacterial species such as MRSA or *clostridium difficile* lay only in the future.

It falls to scientists and technologists to address some of these challenges. One approach lies in miniaturisation of some of the devices which figure in our daily life. Computers, TV's, radios and many other electronic appliances are today but a fraction of their former size. Miniaturisation is a means of reducing manufacturing costs and also, of raw materials, some of which can no longer be considered as inexhaustible. Side by side with this trend, we have another, which is to use less expensive bulk materials, often with inferior properties, while upgrading their performance with more sophisticated surface treatments. Superimposed on these trends is another one – a requirement, increasingly enforced by legislation – that a manufactured product be readily recyclable, in part to reduce the frighteningly large volumes of domestic waste, and in part to allow the economic recovery of metals and other materials which might otherwise be in short supply.

Against this background, we can now consider the issues of "bioactive" metals and surfaces containing such metals, and how best surface engineering can create products incorporating this as a feature. The studies of Nägeli, who formalised such bioactivity under the term "oligodynamic effect" will be discussed in this chapter.

As described below, the effect of silver or copper ions, released from solid metals, metals salts or metal-containing complexes for disinfection of water and other liquids, solids or textiles, has been known and used for decades, even millennia. Silver especially, with its potential to destroy bacteria, viruses and fungal growths, even at very low ionic concentrations, has repeatedly proven its great value. Here we have a substance which is relatively harmless to Man and other mammals, yet with a proven ability to destroy over 650 varieties of bacteria, yeast growths, moulds and fungi (1b). Other metals such as copper show similar behaviour (102) but not to the same extent.

The growing threat of antibiotic resistant species was noted above. Thus in 1948, barely 20% of *staphylococcus* species were penicillin resistant. Ten years later, the figure had risen to 80% (159). Today, annual death rates in the UK alone are in the order of 10,000 or so in hospitals alone, with comparable figures in many other countries. The extent of similar illnesses or fatalities in Care Homes, catering establishments both large and small, can only be guessed at. To these problems we can add that of legionnaires disease, a species of bacteria which thrive wherever warm (but not hot) water is relatively stagnant, notably in institutional hot water systems, but perhaps most lethally in air conditioning plants where evaporative cooling is used. There are too many cases where innocent passers-by in the streets below, were infected by water mist or spray from such installations mounted on the roof of an office block. Invariably, the cause is the same. Maintenance of the equipment has not been carried out, *legionella* growth proliferates and the deadly water spray is dispersed on the wind, travelling many hundreds of meters. The survival rate for humans so infected is of order 50%.

Against this background, silver and, to a lesser extent, copper, by contrast with conventional antibiotics, by virtue of their broad-spectrum bactericidal effects, have emerged from decades of relative obscurity and are today very much in the spotlight. Only occasionally do bacteria and related species prove immune to the action of silver or copper (90, 163). An exhaustive compilation of the prophylactic effects of silver, and the literature reporting this, was provided by Decker (90), much of which is cited in this book. There is yet another piece in this complex jigsaw. In recent years, in all advanced economies, there has been a huge investment in pharmacological research, largely funded by the major drug companies, desperate for new pharmaceutical products to replace those whose patent protection has expired. Opposing this, one might say, is the growing body of legislation which demands exhaustive clinical trials before a new drug can be brought to market. Both in terms of the necessary time for such trials, as well as the huge expense involved, a very significant barrier has been created, which is something of a disincentive to research and development in the field. The prophylactic effect of silver appears to fall outside this legislation as it is presently framed, which might be seen as a plus point. Against that, however, as this book itself demonstrates, the oligodynamic effects described here, with their long history, make it difficult, though not impossible, to file a patent which escapes the Scylla and Charybdis of "prior art" and "obviousness".

Yet another trend reported from every developed economy, is the growth in the proportion of GNP devoted to medical provision. Typically, in most European nations, this is of order 10%. Many economists view this with some alarm. As life-expectancy increases, few see any scenario other than one in which this figure will grow. Then too, in terms of future developments in pharmacology, there are fears that the low-hanging fruit has now been harvested and that advances in drug discovery and other medical procedures will drive up the cost of medical care. Under such budgetary pressures, it is clear that if better use can be made of more modestly-priced medical approaches such as those discussed in this volume, a valuable helping hand would be offered to the management of health expenditure budgets. In this context, it is right to recall the old maxim "Prevention is better than cure". In terms of the present discussion, this should be read as increased emphasis on improved hygiene, whether of work surfaces in kitchens, catering establishments or in the food industry, or in any other location such as hospitals, where open wounds can be found and/or where significant numbers of patients are housed in close proximity to one another, some of them almost certainly carrying bacterial infections, either caught during their stay in hospital, or equally possibly, having previously lain dormant prior to hospital admission. The prevalence of asymptomatic infections in the general public is widely recognised. However hospitals and Care Homes are locations where hygiene becomes specially important, since many inmates have low or otherwise compromised immune systems.

The oligodynamic action of silver and also copper today already plays a significant part in the health sector. Many believe, however, that the potential benefits of these metals is very far from being fully exploited. To achieve this, one thing is clear – the best results will come only if there is good interdisciplinary co-operation. Physicians, life scientists, surface engineers and materials scientists, as well as their technological counterparts – all have a role to play.

The term "oligodynamic" has several times been used, and will be referred to again. It was coined by the biologist Karl Wilhelm von Nägeli (1817 to 1891). He was studying the effect of silver, copper and mercury on a freshwater algae species of the spyrogyra family, and observed a toxic effect on these growths, related to the concentrations of the metals present. His work was published posthumously in 1893 (89). The term he coined has its roots in two Greek words- *oligos* meaning small or few and *dynamis* meaning force. Combined in one, the term oligodynamic means "effective in small quantities". Von Nägeli might be said to have rediscovered this effect in that he found that inserting a piece of silver into a Petri dish containing a culture medium, resulted in a growth-free zone as shown in Fig. 1.1 Given that, at the end of this experiment, no detectable silver weight loss was found, von Nägeli postulated some unknown form of radiation at work. Only much later was it ascertained that release of silver ions causing the observed effect was too small to be detected by him..

Thus it was that an effect known and used in ancient times was rediscovered and, in due course, commercialised. Silver, though not as

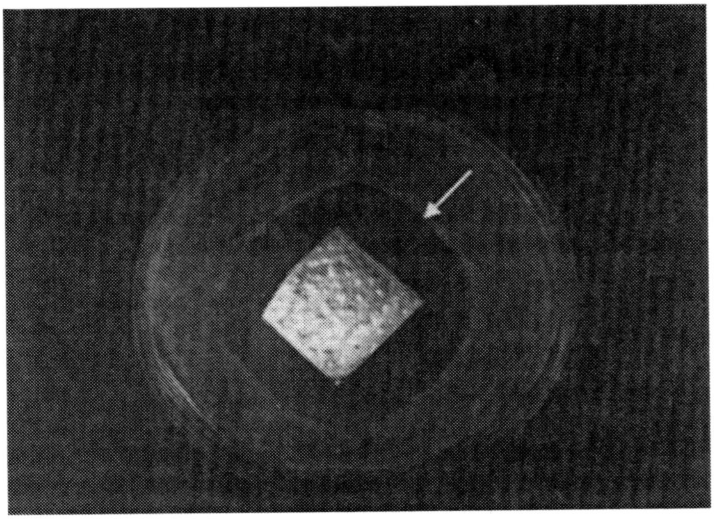

Figure 1.1 Colony-free zone surrounding a piece of metallic silver in a culture medium.

expensive as gold or the platinum metals, is not cheap, and the key to successful exploitation of the oligodynamic action of silver is creating a system whereby a controlled release of silver ions can be achieved at a suitable rate, but without locking up an economically unrealistic amount of the metal. Electroplating and other forms of coating, as we shall see, affords one means of doing this, allowing thin layers of silver to be readily deposited on a wide range of substrates, quickly and efficiently and, most importantly, at tightly controlled thicknesses ranging from microns to millimeters.

1.4 THE PREVALENCE OF BACTERIAL INFECTIONS

That hospitals today are hotbeds of infection, in many cases fatal, is not in doubt. The most common such hospital-acquired infections are those of the respiratory and urinary systems and blood poisoning (2). The German Association of Hospital Hygienists estimates that some 600,000 to 800,000 such infections occur annually in that country alone, of which 20 to 40,000 are fatal (36, 168). Fig 1.2 illustrates this, it is seen that the category "sepsis" is third from top.

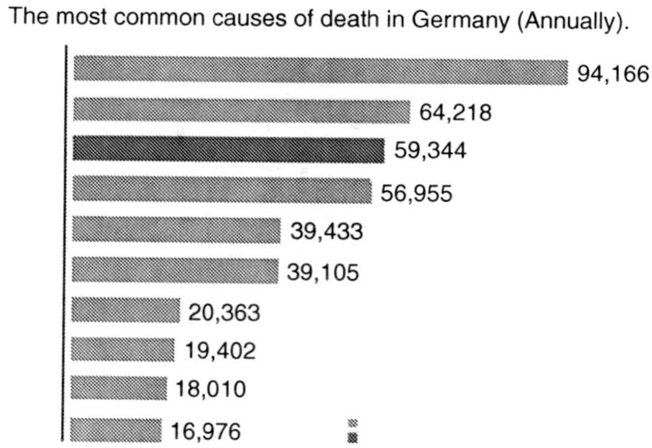

Figure 1.2 Most common cause of death in Germany (168). From top to bottom: coronary (chronic); heart attack; sepsis; coronary (other); stroke; lung cancer; liver cancer; chronic bronchitis; breast cancer; diabetes.

Between 1995 and 2001, according to research at the Paul Ehrlich Institute, the proportion of Methicillin-resistant variants of *staphyloccocus aureus* tested, rose from 8% to 20%.

In recent years, the Japanese in particular, have been quick to recognise the potential uses of silver, in terms of its oligodynamic behaviour. In 1996, wide-scale infections (mainly due to *Escherichia coli 0–157*) were headline news, and largely attributed to poor hygiene in food preparation (3). As a result, numerous products used in food preparation or handling were then provided with silver or silver-alloy surfaces. In the USA, around the same time, a similar increase in the use of silver was noted, mainly in the context of wound dressings (4). In Germany, apart from the established use of silver, electroplated copper-tin alloys were promoted as a means of preventing the spread of infections, largely for equipment and sanitary installations in hospitals. One such proprietary range of alloys is known as "Miralloy". Its bactericidal efficacy is seen in Table 2.1 (a) and (b) and sends a signal that silver might not invariably be the most effective oligodynamic species.

Silver is used as a bactericidal agent in many different forms, including as massive metal, as thin cladding or coatings (which can be deposited by electroplating or by gas-phase deposition, as well as in soluble form or as nanoparticles incorporated in paints, metals, in glassy or ceramic matrices (as so-called nanocomposites).

Numerous products where hygiene is important, such as in food handling and preparation, have been modified so as to incorporate silver or its alloys. In the USA, there has been a clear trend where silver is increasingly used, especially in wound dressings (4). Miralloy, mentioned above, is deposited by electroplating and used to repress infections in hospitals, for example on sanitary ware, where it has been patented (91,92).

The oligodynamic metals can be incorporated into a product in many ways. Most obviously, this is by application of a continuous thin coating. Such coatings can be formed by electroplating, by electroless plating or by deposition in vacuum chambers or from the gas phase. For thicker films, mechanical cladding is a further option. In all of the above cases, not only can pure silver or copper be laid down, but equally, a wide range of alloys of these metals. Another approach is the incorporation of finely-divided metal or alloy particles or powder, in a vehicle such as paint, glass, ceramic or even another metal. The term nanocomposite is generally applied to such materials. Then too, metal particles can be incorporated in or on individual fibres of fabricated textiles (woven or non-woven). In all cases, it is necessary that however such oligodynamic metals are incorporated in the substrate, the rate of metal ion release is sufficient to achieve the bactericidal or fungicidal action required and that such a release is maintained over the projected lifetime of the product in question – a well-established process known as controlled release.

As with any technology, means are required, both qualitative and quantitative, to test performance and efficacy of various materials in terms of their bactericidal, viricidal or fungicidal action on surface-residing micro-organisms. Many such test methods involve a study of the colonisation behaviour of bacteria on the surface to be tested, s well as their adhesion behaviour. Effects on growth, cytogenetic action, cytotoxicity and cytocompatibility in presence of silver and other metals must also be studied. By virtue of the many test methods which have been, or are currently used, is it not always easy to make direct comparisons.

One widely-used test method for bactericidal activity is based on agar diffusion tests, but Challenge tests such as JIS L 1902 (1998); XP39; SN 195924 and electrochemical test methods such as impedance or voltammetry have also proved useful (90). Keevil (179) has, and rightly so, in the author's opinion, expressed reservations about the JIS Test methodology. In the first place, by specifying a test temperature of 37°C, it can hardly be said to replicate actual conditions found on work surfaces, door knobs etc. In the second place, it specifies that a sheet of plastic film be applied over the top of the cell cultures. This might well force them into a quasi-anaerobic mode and could encourage a build-up of bacterial decomposition products. A copy of the JIS Test is available on request from the publishers. Other workers have devised their own test conditions. Thus Robine & Boulange-Petermann (181) studied stainless steel, copper and other metallic surfaces soiled by airborne (as aerosols) bacteria, in their case *Enterococcus faecalis*.In line with other workers, they found massive copper to exert the largest bactericidal action. They also noted the effect of Relative Humidity, an increase of which enhanced bactericidal activity, a finding possibly supporting the role of atmospheric corrosion of the copper as part of the overall mechanism.

1.5 HISTORICAL BACKGROUND TO THE OLIGODYNAMIC EFFECTS OF SILVER AND COPPER

1.5.1 An Age-Old Treatment – Almost Forgotten, Now Rediscovered

The medical history of the beneficent effects of silver, and to a lesser extent, copper (and also their alloys) is probably without parallel, as we shall see. For some two thousand years, the benefits of these metals formed part of medical folklore. Advances in medical science in the 19th Century, seems to have overshadowed this folk-wisdom. Then, towards the end of the century, Nägeli carried out the formal studies which confirmed the basis of the folk-wisdom, and coined the word "oligodynamic". Then came the 20th C and the discovery of various new drugs, but none more important than penicillin. Here, it was thought, was a new class of drug – the antibiotics – which offered a cure for all bacterial infections. As such, and understandably, interest in the older approaches, such as the use of silver, faded almost completely. Only some 40 years later, was the spectre of drug-resistant bacterial species, widely recognised and acknowledged by the medical profession. A frantic race developed, in which, as fast as new types of antibiotic were developed, bacterial resistance advanced. By the end of the 20th C, came the frightening recognition of the fact that bacterial species were becoming more common, which were virtually immune to treatment with antibiotics. It is surely no coincidence that – against this terrifying prospect, interest has returned once more to the use of silver and also copper.

1.5.2 Silver as a Prophylactic – Older Than Recorded History

It would appear that Man's knowledge of the medical or prophylactic benefits of silver and also copper, stretches back to the beginnings of recorded history, and even beyond that. The first written account is believed to be that of the Egyptian Gabir ibn Hayan as-Sufi, dated around 950 AD. At very much the same time, the legendary philosopher and physician, Avicenna (960–1037), in Baghdad also described the medical benefits of silver. He appears to have been the first to describe the symptoms of argyrie, a form of silver poisoning, which manifests itself as a blue tinge of the skin and internal organs, due to prolonged ingestion of excessive doses of silver salts or silver-containing proteinaceous substances.

It is also known that, in ancient Egypt, honey was mixed with silver powder and used as a remedy for a range of inflammations. Silver vessels were widely used in antiquity for the storage of water, both domestically and on sea voyagers and here too, it seems probable that this was a deliberate measure to repress growth of micro-organisms, where other materials such as pottery lacked the same effect. Certainly the Phoenicians, the great sea voyagers of ancient times, used silver vessels for water storage on their ships, and silver is used in water storage tanks on board ships even to this day, in order to maintain water quality. Herodotus, the great Roman historian, describing the Persian Emperor Cyrus the Great (539–530 BC), founder of the Persian Empire, reports how he used silver water containers on his military baggage trains. Figure 1.3 shows a silver tankard in the form of a stag's head, dated to the time of Philip II, father of Alexander the Great. (169)

Roman Centurions and high-ranking officials who used silver drinking vessels and utensils, were much less prone to disease than lower-ranking legionaries who used base-metal utensils. It is widely assumed that similar prophylactic effects protect communicants, where the Communion chalice is invariably made of silver.

In more recent times, American migrants on their long cross-continental trek to California, would throw silver or copper coins into their water containers for the same reason, to prevent the water from becoming brackish. In Europe too, up to the 20th C, housewives would place silver coins at the bottom of enamelled steel milk jugs, to extend the shelf life of the milk. The well-known saying that someone is "born with a silver spoon in their mouth" is widely thought to mean that the

Figure 1.3 - See text.

person had an affluent upbringing. In fact it refers to the oligodynamic benefits of the metal.

Following the rediscovery of the bactericidal effects of silver and other metals by von Nägeli as described below, towards the end of the 19th C, silver and silver compounds were employed for treatment of burn injuries and as disinfectants. These more recent applications of silver as an antiseptic is linked to Credes who, in 1884, introduced Credes Antiseptic (powdered silver citrate) and Credes Salve (colloidal silver as a salve). These were recommended for treatment of wounds and skin injuries. Even to this day, use of silver as a prophylactic medium and occasionally, for treatment of gonorrheal eye infections in neonates (much more commonly in his time) using a 1–2% silver nitrate solution, can be traced to the ideas of Credes. Silver (usually as the nitrate) has long been used for treatment of warts and other non-malignant skin growths, sometimes by topical application of solution, sometimes by use of a "styptic pencil" where the silver salt is immobilised in a waxy vehicle. Table 1.1 summarises key dates in the history of medical uses of silver (5).

By 1940, so many reports of the bactericidal and related medical uses of silver had been reported that, in one publication (99), some 500 literature references were cited. In the same way, a published survey

Table 1.1 Historical Summary of the Medical Uses of Silver

4000 BC, Egypt	Earliest craft silver
Alexander the Great	Silver goblets for drinking water disinfection
1884	Credés Prophylactic
1893	Von Nägeli's theory of oligodynamic metals; various practical applications; Höllenstein, silver-protein combination; silver-arsenic compounds for syphilis treatment
1950	Introduction of antibiotics
1965	Silver nitrate compresses
1968	Silver-sulphadiazine for burns
1980	Vitacuff (subcutaneous cuff) catheter coating

in 1956 on the uses of silver for disinfectation of drinking water and mineral waters (100) cited 162 literature references. Silver was also used as a fungicide. Käszmann (182) recounts how, especially in warmer weather, electroplaters plagued by mould growth in their rinse tanks, found that by suspending one or more sheets of silver in the water, such growths could be repressed.

1.6 AND THE LUNATIC FRINGE ?

As everyone involved with conventional medicine knows, there lie beyond this discipline, layers of "alternative therapies" which range from the questionable, to the wholly implausible, not to say disreputable. Internet searches show that various companies offer "colloidal silver", apparently for self-medicating hypochondriacs. What is the effect of such preparations on the intestinal bacterial flora, one can hardly imagine. But it is quite clear that such a market exists and it is difficult to see how it might be regulated.

1.7 EARLY DEVELOPMENTS

One of the earliest examples of water disinfection was developed by Krause in the 1920's. The Katadyn® and Elektro-Katadyn® processes (86) and similar methods used silver-coated or silver-containing substrates as bacteriological filter media in small-scale, mainly portable water purification units. The CD-ROM database includes both papers and patents from the early 20thC.

1.8 COPPER TOO HAS A LONG MEDICAL HISTORY

The medical and prophylactic uses of copper are at least as ancient as those of silver. The ancient Egyptians are known to have used verdigris (copper acetate, made by reacting the metal with vinegar) for treatment of eye infections. A papyrus scroll, the so-called Smith papyrus, dated at 2600 to 2200 BC, notes the use of copper salts for sterilizing drinking water or as wound-healing promoters. The Greeks used copper oxide, mixed with honey (as was silver, see above) as an antiseptic. The fungicidal properties of copper salts have also long been known, exemplified, for example, by the long-established use of "Bordeaux mixture" (principally copper sulphate) sprayed onto vines to prevent mould growth. Traditional wisdom has it that placing a copper coin in a vase containing cut flowers in water, will extend their life. There seem to be several theories to explain this, one of which is that bacterial growth clogs the pores in the stems, by which water is taken up, and that the bactericidal action of the copper suppresses this.

Copper (often as the mineral malachite) is used by fish-keepers and in some fish farms as a molluscide and the sheathing of wooden ships' hulls with copper sheet had several purposes, of which repressing marine growth, both vegetable and molluscular, were some. Copper wire coils are also used as IUD's (interuterine devices) but the role of the metal here is not clear, and some of these devices are made of plastic, containing no copper.

1.9 THE MECHANISM OF OLIGODYNAMIC ACTION

This section, as its title implies, discusses the science underlying oligodynamism. Immersing oneself in the literature, it is clear that the whole field is divided into two, almost completely separate, topics. At the "front end", we are concerned with the physical chemistry, electrochemistry and corrosion of metals and their salts. This is the science describing the means by which metal ions are formed from metallic or inorganic precursors. Once these ions are formed, the bioscientists, microbiologists, bacteriologists and others take over. Outside these two categories, but of no less importance, are the materials scientists, whose task it is to create metal ion-releasing materials. As will be seen, there is an extensive (and growing) literature relating to the science and technology of the subject. Overwhelmingly, this literature relates to the reaction of micro-organisms, when challenged by metal ions. Although this is undoubtedly an interdisciplinary subject, it is the bioscientists who are *primus inter pares*.

Several explanations have been advanced to explain the phenomenon of oligodynamism, but a key finding, reported independently by Duerr (6) and Acel (7) was the observation that it is silver ions in solution, which exert the bactericidal action, that is to say not the massive metal. However credit should also be given to Ficker (cited by Grumbach in 1949) who, back in 1898 ascribed the oligodynamic action to the presence of metal ions, A similar conclusion was also reached by Spiro (cited by Herzberg in 1923) (135).

A more detailed study of the antibacterial activity of silver was undertaken by Gibbard (8). He reported that metallic silver surfaces exhibited no such activity and only after the action of an oxidising acid on the metal surface, was a bactericidal action shown. This finding supported the work of Duerr, who found no activity on a silver surface which had been reduced with hydrogen.

Based on these findings, a better understanding of oligodynamism was reached using electrochemical methods, where metallic silver was anodically dissolved in physiological salt solutions (9) and this led to the key finding that it is the presence of silver oxide(s) that provides the key to subsequent bactericidal action. It is well known that silver oxide (Ag_2O) has a measurable solubility. In the presence of this compound, the concentration of silver ions is a function of the solubility product of the oxide. A general definition of solubility product of the compound AB is:

$$L_{AB} = [A^{n+}] \times [B^{m-}] \qquad (1.1)$$

where it is assumed that AB dissociates into cations and anions of whatever valency. The solubility product, L is temperature dependent. In the case of silver oxide, L_{Ag_2O} is 10^{-8} ($mol^2 \times l^2$), and the concentration of silver ions will thus be very small. In parenthesis, it might be noted that, as a general principle, if the concentration of either anion or cation in solution is determined by addition of either of these from some other source, that will drive down the concentration of the other species, so that the validity of Eq. (1.1) is maintained. In the case of silver, therefore, we have:

$$Ag_2O + H_2O <===> 2\ AgOH \qquad (1.2)$$

$$\text{and: } AgOH <===> Ag^+ + OH^- \qquad (1.3)$$

$$\text{or: } Ag_2O + H_2O <===> 2Ag^+ + 2OH^- \qquad (1.4)$$

The latter equation predicts an increase in pH when an oxidised silver surface is immersed in water, and this has been experimentally confirmed (9). We thus see that the release of silver ions from the parent metal is contingent on the presence of an oxidising agent, which can be oxygen from air, or this gas dissolved in aqueous media. Given a value for the saturation concentration of oxygen in water, the maximum concentration of silver ions can be calculated (Table 1.2, (90)).

Silver ions liberated from silver oxide into aqueous solution can react with various species commonly found in water, to form silver carbonate, hydroxide or oxide. It follows that the theoretical values shown in Table 1.2 are in many cases not attained in practice (6).

The bactericidal activity of a silver foil, after 30 mins immersion in a 30% hydrogen peroxide solution, attacking the species *Pseudomonas aeruginosa 856*, is shown in Fig. 1.4, with – as a control – silver not artificially oxidised. The latter does show a slight activity in comparison with a totally silver-free sample, but even so, does not impede growth of the culture (Fig 1.4(a)). By contrast, the H_2O_2 oxidised foil, as also a pressed pellet of silver chloride, appear to repress growth of the culture completely (Fig 1.4(b)) (90).

Thus it is seen that for the silver to exhibit antibacterial activity, suffcient silver ions must be present in a moisture-containing environment. Silver ions can be generated not only from the metal oxide or its salts, but also by corrosion based on a local galvanic couple or anodic dissolution of the metal. Silver ions demonstrate a bactericidal activity at concentrations as low as 10^{-9}M (ca 0.1µg/l) (12). According to older work by Fromherz (13), a bactericidal action is seen above a threshold value of 10^{-11}M. In fact, even the more insoluble silver salts provide a silver ion concentration above this value, for example silver iodide. Only the sulphide, with its uniquely high solubility product of 10^{-51}, shows no oligodynamic activity. Berger et al (170) reported the minimum inhibitory and bactericidal concentrations for 16 types of

Table 1.2 Maximum attainable silver ion concentrations in oxygen-saturated water in presence of massive silver, as function of temperature (theoretical values)

	Sol'y of oxygen in water (atm. Press) gm/l	Sol'y of Oxygen in water (atm. Press.) M	Silver ion conc'n (M)	Silver Ion conc'n (gm/l)
20 °C	0.044	2.75×10^{-3}	1.10×10^{-2}	1.19
40 °C	0.033	2.06×10^{-3}	8.25×10^{-3}	0.89

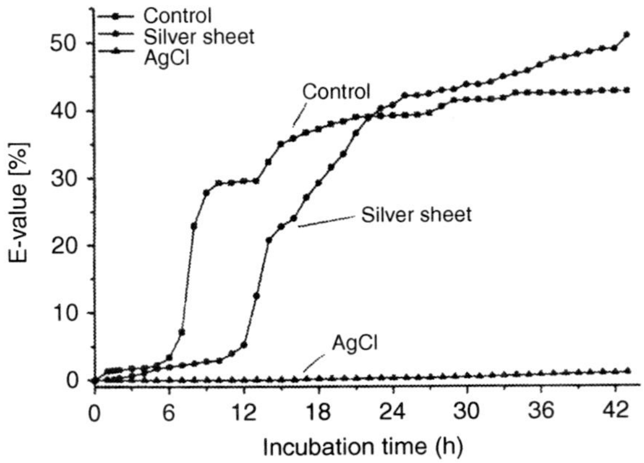

Figure 1.4(a) Efficacy of a silver foil (total area = 150.5mm^2) and a pressed silver chloride pellet (area = 180.6mm^2) challenging *Pseudomonas aeruginosa 856* (Initial colony count 4.6×10^2 cfu.ml^{-1}). Incubation volume 6.1ml (90).

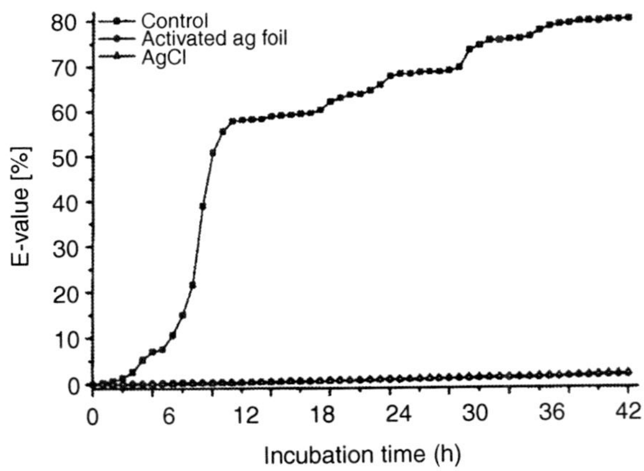

Figure 1.4(b) As above, but silver foil activated with H_2O_2. Initial colony count 6.7×10^4 cfu.ml^{-1} Incubation volume as before.

bacteria, using electrolytic dissolution to form known amounts of silver ion. Their results show that all species studied by them were growth-inhibited at 1.25µg/ml silver ion, while at 10.05µg/ml, all colonies had died. Table 1.3 lists data for a range of sparingly-soluble silver species.

Not all of the above species are in fact oligodynamically useful, either in terms of their cytotoxicity or the silver ion amounts released. In the case of the most soluble species such as silver nitrate or perchlorate, which are completely ionised, the formal concentration is the same as

Table 1.3 Solubility of some silver salts (138–140)

Silver Species	Formula	M. Pt (°C)	Sol'y Prod. K_L (M)	Temp. (°C)	Silver ion conc'n (M)	Silver ion conc'n (mg/l)
Bromide	AgBr	430	7.7.10E-13	25	8.8.10E-7	0.09
Carbonate	Ag_2CO_3	200	6.2.10E-12	25	2.3.10E-4	25.00
Chloride	AgCl	455	1.6.10E-12	25	1.3.10E-5	1.36
Cyanide	AgCN	320	2.2.10E-12	25	1.4.10E-6	0.15
Hydroxide	AgOH	250	2.0.10E-8	25	1.4.10E-4	15.25
Iodide	AgI	558	1.5.10E-16	20	1.2.10E-8	1.32.10E-3
Oxide	Ag_2O	250	3.0.10E-12	20	1.8.10E-4	19.60
Sulphate	Ag_2SO_4	652	7.0.10E-8	20	5.2.10E-3	560
Sulphide	Ag_2S	825	6.3.10E-51	25	1.2.10E-17	2.5.10E-12

that of the silver ions released.. It should also be noted that, by virtue of the penetration of silver ions into the cell structure, and adsorption of the ions on the bacterial matter, as well as onto the walls of the containing vessel, a decrease in the concentration of free silver ions in solution is observed. Thus it may be that the concentration of these ions falls below that required for cytotoxicity (an actual example of this, based on military use of silver-based water purification tablets, is quoted in a subsequent chapter).

When using the more sparingly-soluble silver compounds, a longer contact time with the liquid to be disinfected may be necessary, in that the silver ions are only slowly released. By contrast with the more soluble silver salts, a much slower, demand-related release of silver ions can be found, from the more sparingly soluble salts, where such delayed diffusion from the solid is governed by the Law of Mass Action. These thereby act as controlled-release systems.

Silver chloride is a non-toxic species which, with solubility of 1.26×10^{-5} M, provides sufficient silver ions to act oligodynamically. At higher pH values (9 to 12), the rate of silver ion release increases markedly, as shown in Fig. 1.5 (90)

Almost all silver compounds are photosensitive. Exposed to light, silver chloride decomposes to metallic silver and chlorine gas, which acts as a bactericide. This liberation of chlorine is offered as an explanation for the often conflicting results found in microbiological studies (90). Silver sulphadiazine (AgSD) is a sparingly soluble silver compound, used since 1965 as a cream for prevention of infections or treatment of infected burns. A range of protein-containing silver compounds are characterised by their relatively good solubility. The proteinaceous component acts as a colloid stabiliser, thus preventing

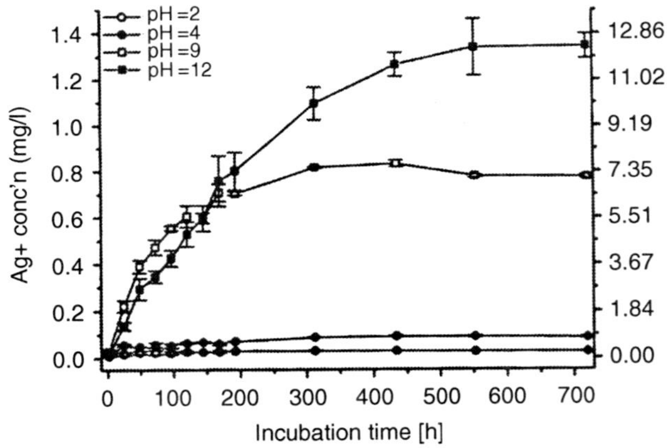

Figure 1.5 pH dependence of silver ion release from pressed silver chloride pellets (90).

agglomeration during drying. Examples of protein-solubilised silver compounds include "protein-silver", silver protein acetyltannate and silver vitellinate. Table 1.4 lists silver-containing medicaments licensed for use in Germany, with indications for their use.

The microbial activity of silver ion containing compounds such as the above, depends on the binding of the metal ion to the electron donor group on the biological molecules, which contain sulphur, oxygen or nitrogen. The silver ions act by exchanging with other key metal ions such as Ca^{2+} or Zn^{2+} (10, 11). Organic compounds with sulphur-containing groups (cystine, thioglycolate, glutathione) react immediately with silver ions, and immobilise these, thereby nullifying the antimicrobial action of the silver ions (164, 165). In recent years, the cytotoxic action of silver ions has been even better understood (14–17, 141–147). Decker (90) has summarised it as follows:

As a function of the metal ion concentration, two different mechanisms should be distinguished (141). At low metal ion concentrations, a physical adsorption of the metal ion on the cell wall or cell membrane of the micro-organism takes place. This process constitutes a bacteriostatic action which simply represses cell growth. At higher metal ion concentrations, a bactericidal action is seen, which is associated with permeation of the metal ions through the cell wall into its interior. Various mechanisms result in damage to the micro-organism including a blocking action of some of the structural and functional proteins, as well as interactions with the nucleic acids (18, 147).

Table 1.4 Silver-containing pharmaceutical preparations authorised in Germany as of 1996 (Red List)

Silver compound	Concentration & Form	Brand Name	Indications
2-aminoethyl-dihydrogenphosphate, mono silver salt	3% silver salve or 5% gel or powder (external use)	Dermazellon®	Prevention of infection, treatment of wounds & burns; after skin transplants
metheneamine silver nitrate	2% powder (external)	SilverPowder	External injury, cuts, injection injuries, wound infection, umbilical healing.
Colloidal silver	0.25% sol'n (combined with plant extracts camomile flowers, peppermint leaves)	Gastractin®	Acute & sub-acute gastritis, acute & chronic enteritis, summer diahorrhea, infant diahorrhea
Metallic silver	0.2% salve (with urea & magnesium chloride) external use	Green Salve "Schmidt-N"	Prevention of infection, treatment of wounds & burns, after skin transplants.
Silver nitrate	1% Sol'n. Eye drops.	Mova Nitrate Pipette®	Prevention of blenorrhea (Credes prophylaxis)
Silver nitrate	As solid (styptic pencil). External use after moistening	Höllenstein Stick	Wounds, warts
Silver sulphate	Magaldrat & Simethicon as active ingredients. Silver sulphate declared under "other constituents" (internal use)	Riopan®Gel	Acid-related stomach conditions. Ulcus ventriculi & ulcus duodeni
Sulfadiazine Mono silver salt	1% cream (external use)	Demazin®	Preventing infection of burns, warts, sores, ulcers Cauterising
	1% cream (external use)	Flammazine®	Prophylaxis & therapy of wound infections after Burns; cauterising

The effect on the microbial proteins is associated with the high affinity of silver (also copper) ions to electron-rich functional groups such as hydroxyl-, amino-, carboxyl- and especially sulphydryl- which are necessary to the functioning of enzymes and structural proteins. In consequence, defects occur in the cell membrane and cell walls, with a blocking of enzymatic action. In particular, enzymes associated with energy-releasing fermentation and respiratory chains containing sulphydryl groups are affected.. Damage even to a single enzyme results in a breakdown of the entire fermentation mechanism with a collapse of the energy-matter interaction (142, 143). Thus an inactivation caused by silver ions can be triggered using silver nitrate, as shown in the following scheme;

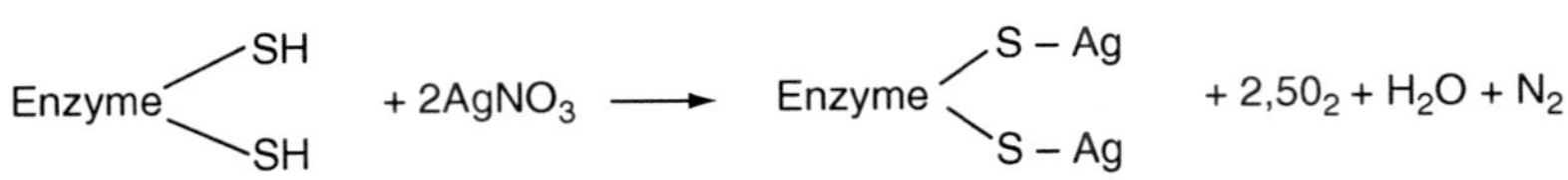

The strongest blocking effect on the respiratory process due to silver ions is that between b-cytochromes and cytochrome a_2. To a lesser extent, is that between NADH (nicotinamide-adenine-dinucleotide) or succinatedehydrogenase and flavoprotein on the side of the substrate entry (144).

The deactivation of bacteria can also be attributed to complex formation between the metal ions and nucleic acids. The affinity for silver ions to DNA (desoxyribonucleic acid) is some 40 times higher than that to RNA (ribonucleic acid) (145). The basic bonds associated with hydrogen-bonds undergo a change in spatial structure as hydrogen atoms are exchanged for those of silver. This is due to the difference between the lengths of N-H and N-Ag bonds. In addition to breakdowns caused by enzyme blocking in the energy-matter exchange mechanism, strain or indeed rupture of the Watson-Crick double helix structure, can occur, thereby affecting the protein biosynthesis process. Renewed microbial growth is observed when a silver atom becomes associated with an 80-base-paired double-helix structure (145).

Gram-positive* micro-organisms are characterised by a homogeneous cell wall of 10 to 80nm thickness, Gram-negative* cells, by contrast, have a dual-layer membrane with total thickness of 9–11nm. This results

* A sub-classification of micro-organisms was established by the Danish physician H C J Gram, in terms of their response to staining. Gram-positive stain dark blue, Gram-negative stain red.

in a more rapid penetration of metal ions into the cell interior in the case of gram-negative bacteria which are thus more susceptible to the action of metal ions than gram-positive types (132, 133, 148–150) and the lowest sensitivity to silver ions is found in *enterococcus* (132, 151). Using silver chloride aerosols, anthrax spores were killed off within 15 to 30 minutes (152).

The efficacy of silver sulphadiazine aginst yeast growths such as *candida albicans* and various fungii (incl. *aspergillus fumigatus* and *aspergillus flavus*) which present a serious challenge to patients with large-area burns, was clinically confirmed (155).

In addition to the bactericidal action of silver ions, their viricidal action against influenza- and mumps-type viruses has also been established (149). Thus the *Herpes-Simplex* virus was found to have been inactivated by application of silver nitrate, ascribed to a blocking of the –SH groups on the capsid membrane (156). In other work, the efficacy of silver sulphadiazine against *Plasmodium berghi* was confirmed in tests using CF-1 strain mice, both when orally or subcutaneously applied (157). In-vitro tests showed that *Amoeba* was killed off by silver nitrate or silver uracil*. (158). In some instances, a reduced resistance to silver ions has been reported. The origins of such a resistance has been explained in terms of protein formation within the cell. These complex the silver ion and thus, by reducing its activity, make it less toxic. Physical effects too, can weaken the anti-microbial action of the metal ions, for example by formation of a slimy layer.

A resistance to the effect of silver ions of certain human pathogenic bacteria has been observed in the case of certain micro-organisms (90) such as *Enterobacter cloacae*, *Escherichia coli RI* (possibly by enhanced production of hydrogen sulphide and the resulting precipitation of silver sulphide). Also *Escherichia coli* (decreased silver uptake), *Pseudomonas* (reduction of silver ions to elemental silver), *Pseudomonas maltophila* (reduction of silver ions to the metal), *Pseudomonas stutzeri AG 259* (non-specific binding of the silver ions to negatively-charged grouping within the cell), *salmonella typhimuriem* (reduced uptake of silver) and *Citrobacter freundii* and *Proteus mirabileis* (163).

Hydrogen sulphide producing bacteria such as *Salmonella* or *Proteus* also exhibit a significant resistance to the effect of silver ions, again explained in terms of silver precipitation as the sulphide. However by using a rapid releasing source of silver ions or silver chloride

---

* 1:1 combination of silver nitrate and uracil (2,4-dioxypyrimidine) with a solubility of 2.36 mg/l.

containing solutions, such resistance can be reduced (132). Another species exhibiting a reduced susceptibility to attack by metal ions is *Klebsiella pneumoniae*, with a polysaccharide capsule. The slimy outer coating of the cell creates a physical barrier to passage of metal ions (153). In other work (154), an increased resistance to the action of silver ions has been correlated with the lipid content of bacteria.

The oligodynamic action of various metals differs markedly. In the literature can be found numerous rankings of the efficacy of different metals, in many cases apparently contradictory. Thus, in studies of metal toxicity towards fungii, this was found to increase with increasing atomic weight, with silver and osmium being found as most effective (19). Martin (20), quotes the following ranking in terms of fungicidal action:

$$Ag > Hg > Cu > Cd > Cr > Ni > Pb > Co > Co > Zn > Fe > Cu$$

He also established correlations with temperature, pH and exposure time. Hess and Reiter (21) published the following ranking, based on observations of the haemolytic action of various metals, in terms of their oligodynamic efficacy:

$$Cu > Cd > Pb > Ag > Ni > Mo > Sn > Al.$$

Thiele and Wolf (cited in 1995 by Walhäuser (131)) list the following metals in order of decreasing activity:

$$Cd > Ag > \text{brass} > Cu > Hg$$

Hauschild (cited in 1968 by Praetorius (132)) established the order of activity as:

$$Cu > Hg > Ag > Pb > Sn > Zn > Al > Fe > Mn > Pt$$

while Tamann & Rienäcker (cited in 1932 by Neisser & Eichbaum (133)) reported the ranking:

$$Hg > Cu > Ni > Zn > Cd > Ag$$

In 1974, Golubovich et al. (162) reached a somewhat different conclusion. Silver, they assert, exhibits the highest toxicity to micro-organisms, while at the same time having the lowest toxicity to mammalian cells. These authors quote the following toxicity ranking to micro-organisms:

Ag > Hg > Cu > Cd > Cr > Pb > Co > Au > Zn > Fe > Mn > Mo > Sn

Vashkov (167) adopted a somewhat different approach, classifying metals into five groups in terms of their oligodynamic action. These were:

(1) copper, silver, gold
(2) zinc, cadmium
(3) indium
(4) tin, lead
(5) arsenic, antimony, bismuth

There could be many explanations for the apparent discrepancies in these various rankings, whether in terms of the experimental techniques employed or the micro-organisms selected as test species. The five groupings of Vashkov, cited above, would be seen by any chemist as a trivially obvious expression of the Periodic Table. Tamman & Rienäcker (cited by Neisser & Eichbaum in (133)) sought to establish a unique correlation between the oligodynamic activity of the metal and its standard electrode potential. In terms of the various rankings reported above, and after eliminating those metals with unacceptable toxicity (to humans), copper and silver, in metallic form or as their salts, together with a handful of alloys containing these metals, appear to be the outstanding candidates for every-day use as bactericides, viricides and fungicides.

There are other reports on formation of biocidal coatings which, in broad terms, appear to rest on oligodynamic action, but in a manner which is not quite clear. Thus Rudnev, Tyrina et al (183) formed relatively thick anodic coatings on titanium or aluminium, using a process known as micro-arc oxidation or plasma anodising. When species such as sodium hexametaphosphate (HMP) and salts of zinc, magnesium, nickel or manganese were added to the electrolyte, the coating acquired biocidal properties, using a test method where fish-peptone agar medium was sprinkled with suspended bacteria isolated from sea water (from Rynda, in the Japanese sea) Results are not easy to understand, and coatings based on HMP + zinc appeared to have the strongest bactericidal action, while addition of nickel, and or manganese salts to the electrolyte reduced the bactericidal effect.

1.10 NON-METALLIC BIOCIDAL SPECIES & SURFACES

A large number of chemically-active biocidal species have been used in the past and still are today, from the simplest chemicals such as chlorine, chlorine dioxide or ozone to a wide range of organic compounds. Some are mentioned in the text. Ultra-violet light is also used, sometimes in combination with ozone. All of these have their drawbacks. A quite different approach has recently been developed at MIT by Klibanov et al. (177). This utilises a polymer coating which forms needle-like protrusions formed of carbon structures, some 12 atoms high. It is claimed that these "spikes" rupture the viral envelopes and bacterial cell membranes. The coating can be applied as one would a paint. As described, the structure of the surface appears similar to so-called superhydrophobic surfaces, where similar sharp protrusions prevent the surface from being wetted, and one might speculate that such non-wettability might play a role here too.

1.11 TOXICOLOGY OF SILVER AND COPPER

Silver and its compounds are far less toxic to humans, both in acute and chronic terms, than most of the other heavy metals and their compounds. This is due in part, to the very low resorption of the metal by the digestive tract and also to the ease and rapidity with which the metal and its salts are converted to highly insoluble species such as silver chloride, silver albuminate and binding to metal ornithins within the organism. A typical figure for the daily dietary uptake of silver lies in the range 1 to 88μg/day (47, 136). Of this, only some 10% is absorbed by the body, while 90% is excreted (47).

To the extent that water-soluble silver compounds such as silver nitrate (Höllenstein) or silver fluoride are hazardous, this is due to an etching effect of the anions and nothing to do with the silver cations. Applied to the human eye, a 10% silver nitrate solution can cause severe burns. It can be the case that, by long-term ingestion of silver (as a silver salt, in colloidal form or as the finely-divided metal), a significant concentration will build up within the body. This condition is known as Argyrosis or Argyrie and the most obvious symptom is a colouring of the skin, either brownish or blue-black, most commonly slate-grey. As might be expected, such pigmentation is most marked in those areas of the skin exposed to daylight. Long-term use of silver colloids such as silver-protein-acetyl tannate (silver albuminate) (7.5% to 8.5% colloidal silver) can result in Argyrosis. A value of 0.9g has been cited as a typical dose resulting in argyrosis (41). While silver accretion in the skin may have no adverse consequences other than purely cosmetic, a build-up of the metal in the eye is known to damage the light/dark adaptation mechanism (48).

Patients with extensive burn injuries are often treated topically with silver-containing preparations. Serum concentrations of 1.1μg/24hr have been recorded in such cases without any associated argyrosis or other adverse symptom (49).

The maximum permitted workplace concentration of airborne silver (MAK) is 0.01mg/m^3. A typical uptake figure for silver from nutrients and drinking water should be less than 0.1mg/day (98).

The question of copper toxicity is more complex, in that – unlike silver – it is an essential trace element, and dietary deficiencies are harmful. However at the other extreme, copper (in soluble form) is toxic at concentrations (approximately) of 140 mg/kg. Another quoted value is 10 gm. Copper poisoning results in nausea, vomiting, diarrhoea and ultimately death. Among several deleterious effects, liver damage is noted. As far as known, all mammals as well as many lower aquatic species are susceptible. Most other metals are toxic in sufficiently large doses or when ingested in a particular way. Ref. 23 provides a compilation of minimum essential daily intake, systemic toxicity, cytotoxicity and allergenic reaction. One might just note here the high allergenic activity of nickel and cobalt, and the toxicity (including carcinogenicity) of chromium, cadmium, nickel and beryllium. The usual modes of ingestion are orally and nasally (as dusts or liquid sprays). In the latter case, nasal cancers have been many times reported. In the context of this book, it might be asked to what extent contacting a metal with the skin can be harmful. The writer has heard of cases where people (presumably already nickel-sensitised) react to contact with stainless steel. Such cases, if indeed they exist, must be extremely rare. But otherwise, the writer is not aware of any adverse reactions from direct skin contact with metals in their massive and unoxidised form.

There are large amounts of readily-available information relating to the biomedical use of metals and alloys (as biomaterials, implants etc) such as (95) as well as their haemocompatibility in medical applications (96) together with numerous studies of the biological effects of dental alloys (97).

A valid question might be where to establish the boundary between the oligodynamic effect of metals (beneficial) and their toxic action.

Paracelsus offers an answer. "The dose determines whether something is toxic or not". Whereas even very small amounts of silver can act to kill micro-organisms, such concentrations (0.01 to 0.1 mg/l) are harmless to the human organism (98). Potable water supplies in large American cities such as Atlanta, Denver, New Orleans and New York typically contain 0.2 to 0.3 mg/l silver which occurs naturally, rather than being deliberately added. No adverse health effects have ever been reported from this (114). There are no known reports of any mutagenic or carcinogenic effects of silver.

1.12 THE ECO-TOXICITY OF SILVER & COPPER

Apart from bacteria, some fish, for instance trout, and other water-resident micro-organisms are much more sensitive to silver salts than humans, other mammals and plants. The German Gelsenkirche List (1972) sets out the silver concentrations which might cause harm or prove toxic, for four categories of species. These were the simplest forms of waterborne life (lower organisms), (N), cold-blooded vertebrates (K), warm-blooded vertebrates (W) and humans (M). Table 1.5 quotes the threshold values where specified.

Silver and its compounds are not currently seen as an environmental problem and discharge of these into water courses ands streams does not call for special measures. However there appears to be something of a contradiction between this widely accepted truth on the one hand, and a ruling from the EPA (US Environmental Protection Agency) on the other. Referring to biological domestic septic tanks, they include silver with the usual "heavy metals" and gold, platinum & palladium, all being rated "very toxic" and prone to "tip" the natural bacteriological equilibrium which exists in such installations (98).

A further contradictory ruling stems from the German Water Protection Agency EAWAG (Eidgenössischen Anstalt f. Wasserversorgung, Abwasserreinigung und Gewässerschutz (121) who ruled that septic

Table 1.5 Threshold Values of Silver Ion Concentration for Various Species (in mg/l)

	N	K	W	M
Silver nitrate	–	0.003–0.04	–	–
Silver salts	0.01–0.03	0.003–0.1	–	0.05 (chronic)

tank installations might be considered as extremely robust, harmed only by chlorine, cyanide and certain specified organic compounds. No specific metal ion concentrations are laid down by them for such plants. Indeed they state "Some studies have indicated that concentrations even as high as 1gm/litre of zinc or copper have not proved harmful to the bacterial flora in septic tank systems (121). Adverse effects of silver in such tanks was only noted at concentrations of 0.5 to 0.8 gm/l (121).

Copper is widely used as a fungicide (in agriculture, horticulture and fish-farming) and as an algicide. There appear to be no regulations restricting its use for such purposes, though its discharge in effluent streams is controlled at values of 10 ppm or less.

1.13 SILVER FOR WATER PURIFICATION

Minimum purity requirements for drinking water quality are laid down in most developed economies. In Germany, for example, Para. 37, Section 1 of the Prevention of Infection Legislation lays down that water for human use must "by its use or consumption not be harmful to human health, and should not contain matter that might cause illness" (German legislation covering spring- and well-water quality, 20th July 2000, p. 1045). Legislation framed one year later, deals with specific substances. Thus in a 100ml sample of potable water, neither species such as *Escherichia coli* nor coliform spores or *enterococcus* should be detectable. Colony counting may be carried out by different methods depending on whether the older method (cultivation on agar gel nutrient) or the newer method laid down in EN ISO 6222, is used. Using the former methods, the following criteria must be met:

Colony count at 22°C:

- 100 per ml at the water mains stopcock of the user
- 20 per ml at a location immediately adjacent to the water treatment plant
- 1000 per ml for small installations

Colony count at 36°C

- 100 per ml at the water mains stopcock of the user

Using the newer method, colony counts at 22°C and 36°C should show no anomalous changes. In order to establish the normal colony count behaviour (by way of calibration), the newer and older test methods should be run side by side for one year.

In addition, German law requires that potable water be not only free from species harmful to health or other chemicals at harmful concentrations, but that it should be of palatable quality. Potable water is not required to be "pure" in the chemical sense – it will invariably contain mineral salts, indeed in their absence, water has a "flat" and unpleasant taste. Water that is unduly soft (low in calcium and magnesium salts) is associated with elevated risk of coronary and circulatory disease. It should be noted also that the presence of micro-organisms, provided their concentration is not unduly high, is tolerated.

It might be mentioned here that bottled mineral water (uncarbonated), sold all over the world, is regularly tested by government health authorities. Especially where such bottles have stood for too long on display shelves, an elevated, if not life-threatening, level of micro-organisms is regularly reported. The explanation offered is usually in terms of the effect of ultra-violet rays in sunlight, probably acting in conjunction with plasticizers, leached out from the plastic bottles.

In terms of municipal water supplies, after purification at a central plant, the water is distributed to consumers via a water mains system. In these pipelines, this water is pressurised, it is usually continuously flowing, it attains the temperature of the soil in which the pipes are buried – perhaps 10°C, and is in total darkness. None of these conditions is conducive to bacterial growth. However once the water reaches the premises where it is used, things are very different. Water may be stored in tanks, sometimes immediately under a roof where in summer, it will become warm. In parts of the domestic pipework, water may be stagnant for weeks on end. Because such premises are heated, water flowing in copper pipes with their excellent thermal conductivity, may attain ambient room temperature. Indeed where hot water and cold water pipes run side by side, even cold water can become warm. In all such cases, as in commercial and industrial premises where cooling towers are often found (usually associated with air-conditioning plant), large volumes of water circulate, often at temperatures warm enough to encourage bacterial growth, but not hot enough to kill potentially harmful micro-organisms. Such conditions are ideal for so-called Legionnaires disease, which is often fatal (104, 105). The World Health Organisation (WHO) have devised a set of wide-ranging recommendations aimed at eliminating such hazards (106).

Thanks to advances in high-resolution cell typing, an unequivocal correlation has many times been established between waterborne infectious species found in hospital water systems and severe hospital-acquired infections of patients. (107). These findings have had major repercussions in terms of epidemiology, preventive medicine and in legal issues. Transmission of hospital-acquired infections from water supplies is not simply limited to drinking of such water, but can also be due to inhalation of contaminated aerosols (*Legionella*, atypical mycobacteria) or by direct or indirect contact with contaminated water, for example during washing of patients or cleaning medical appliances with contaminated rinsewater. A somewhat similar situation can arise during a sea voyage. Surrounded by water, yet without drinking a drop of it – on-board drinking water supplies, especially in warmer climates, call for scrupulous care in maintaining sterility. The same is true for static marine installations such as oil rigs.

In Germany, the law prescribes what chemicals may be used for disinfectation of drinking water supplies and comparable legislation exists in most developed countries. The permitted species are as follows:

(1) chlorine (gaseous), sodium hypochlorite ("bleach liquor"), calcium hypochlorite, chloride of lime, magnesium hypochlorite, chlorine dioxide, ammonia & ammonium salts.
(2) ozone
(3) silver, silver chloride, sodium silver chloride complexes and silver sulphate
(4) physical methods such as UV irradiation

In the current German legislation, silver as a chemical parameter is no longer included. In the 1990 legislation, silver concentrations were specified as 0.01mg/l for normal water supplies with 0.08 mg/l for static water supplies (bottled or canned).

Silver salts (nitrate, sulphate, carbonate) are, however, included in a listing of permitted additives issued in 2001 by the German Umweltbundesamt. They are permitted additives for stored (bottled or canned) water supplies or in small-scale water treatment plants, but are not recommended for systematic use except in emergencies. The permitted concentration is 0.1mg/l silver, with the highest permitted concentration after completion of the purification process being 0.08mg/l (legislation expired 1st June 2006).

In the USA or Japan, silver-based water treatment plants are permitted, whereas in Germany they are not. As part of the history and problems relating to silver concentration limits in potable water in German (and subsequently EU) legislation, the following comments by Roser (114) from the Annual Report of the (German) Precious Metals Association are cited:

Silver additions were used in Germany for preservation of non-alcoholic drinks back in 1929 and a limiting concentration of 1mg/l was specified for these. After many years of further study, especially in Switzerland, this value was revised to 0.2mg/l which was deemed sufficient for maintaining sterility. This concentration was often applied in other, but related applications. In 1959, German legislation reduced the permitted limit to 0.1mg/l, this being deemed sufficient to maintain an adequate level of sterility.

In a subsequent proposal for EU legislation, a maximum permitted concentration of 0.01mg/l was indicated (with no Guideline). In the associated classification, silver was totally excluded as a permitted bactericide, on the basis of its alleged toxicity. However after technical input from various sources, including Messrs Degussa, the metal was re-classified from "toxic" to "undesirable" and a maximum permitted concentration of 80µg/l was agreed for bactericidal applications. To set this in context, the following table shows maximum permitted concentrations and Guideline values for a range of substances.

The logic underlying these values is not easy to understand. Thus it will be noted that the ZHK for silver, at 10µg/l, is a fifth of the permitted values for far more toxic species such as arsenic, cyanide or lead. The German Health Ministry which prescribed these values, appears to have used the analytical limit of detection, for silver. In other words, no detectable level of silver (using the prescribed analytical techniques) is permitted. In the case of arsenic, cyanide or lead, for example, the permitted maximum concentrations appear to have been set on the basis

Table 1.6 EU Permitted Drinking Water Specifications (in ppm)

Species	RZ	ZHK	Comments
NO_3	25,000	50,000	< 25, 000 for infant feeding
NO_2	——	100	
P_2O_5	400	5,000	
NH_4	50	500	
Fe	50	200	
Cu	100	3000	ZHK after 12 hrs static in pipe
Zn	100	5,000	ZHK after 12 hrs static in pipe
F-	——	1500	
Ag	——	10	After treatment with Ag, ZHK = 80

Notes: ZHK – Max. permitted concentration. RZ – Guideline concentration CN, Pb have ZKK value 50 ppm (no RZ value).

of their ADI values (Acceptable Daily Intake). In the case of arsenic, as a known carcinogen (115), this is clearly critical. In the case of silver, by contrast, it is hard to escape the conclusion that the RHZ value was selected on a purely arbitrary basis. Given the growing use of silver-based water purification systems, the present situation is far from satisfactory.

It is well known that the potable water supplies of several major US cities (Atlanta, Denver, New Orleans, New York) have a silver ion concentration in the range 200–300μg/l. This occurs naturally in the subterranean water sources on which these cities draw, and not as a result of any deliberate dosing of the supply. There are no reports of any kind indicating any adverse effects in the populations served by these supplies (101) and it is clear that the authorities involved have satisfied themselves as to the harmlessness of this situation.

In 1949, in the aftermath of the war, the authorities responsible for the German town of Clausthal-Zellerfeld recognised a need to enhance the quality of the mains water supply and decided to use silver-dosing, in order to avoid the unpleasant taste of chlorine (117). At least ten other German towns or communities followed suit, in addition to numerous privately-operated water systems. In these cases, a level of 20 to 50μg/l silver ion was specified, and at no time then or since, was there any questioning of these measures.

There have been many studies of silver in the human body (118–120). Thus, West et al (120) described how silver is taken up by the liver, and then excreted in the faeces. To date, there have been no scientific reports suggesting that silver might accrete in any human organ, with ultimately damaging consequences.

We must remain hopeful that the ZHK value for silver, as laid down in the EU Potable Water Guidelines, will be revised so as to permit a sufficient concentration of order 100μg/l to allow an unrestricted disinfection of drinking water. Such a value would, inter alia, align silver with the value for lead, a species known to be toxic. Indeed, were such a re-alignment not to be implemented, it seems quite possible to contemplate a prohibition on use of silver cutlery and other dishes, since the use of these quite possibly results in ingestion of comparable amounts of silver.

In order to sterilise small volumes of water, either domestically or during travelling, boiling is the default procedure. To achieve sterility, 15 to 20 minutes at 100°C is normally recommended. Another alternative is filtration (e.g using micro-ceramic or fibreglass filters). A proprietary product, Micropur® is also available in tablet, powder or liquid

Table 1.7 The three most commonly encountered forms of contamination in impure drinking water. (103)

Species	Known types	Size range
Bacteria	Escherichia coli, Salmonella, Cholera	0.2–5 μm
Viruses	Hepatitis A, Norwalk-Virus, Rota-Virus, Polio-Virus	~0.02–0.2 μm
Protozoa	Amoebic dysentery. Giardia, Lamblia, cryptosporidia	1–15 μm

form, for purification and sterilisation of water of doubtful purity. This product is based on a silver chloride – sodium chloride complex, where the former is present at around 1%. Formation of a silver ion complex such as dichloroargentate $[AgCl_2]^-$ or the tetrachloroargentate $[AgCl_4]^{3-}$ increases the concentration of available silver.

In order to remove the most commonly found bacteria in water, a micro-filter with pore size < 0.3μm is necessary (Table 1.7).

In many developing counties, cholera, Hepatitis A and Typhus are widespread (103). Use of silver, preferably in ionic form, is a long-accepted means of combating these and other infections. The slow release of the bactericidal ions ensures an efficacy lasting for up to 6 months, during which cell growth is inhibited.

Numerous studies, carried out in the 1950's at the Institute of Hygiene of Vienna University, established and confirmed the correlation between silver ion concentration and the bactericidal or viricidal action of silver. These studies suggested that 50μg/l would normally provide an adequate concentration for disinfection of water, though in some conditions, for example where only a few hours contact were possible, or where a higher concentration of contaminant was present in the untreated water, then 80μg/l was preferable. Similar concentrations are also advisable where water is stored for prolonged periods without risk of deterioration, for example in ship's drinking water tanks or where water is stored for use in emergencies.

One means of dispensing silver ions for water treatment, is to use anodic dissolution of massive silver anodes. Knowing the current passed and the time involved, silver ions can be very accurately metered into the water (Faraday's Law). The addition of solid silver compounds has previously been noted (103). The metal in these cases is usually present as a complex. It is usually assumed, in such cases, that the water has previously been filtered and is free of suspended matter or turbidity. Silver may also be incorporated in the filter itself. By virtue of its simplicity and ease of operation, silver as a water disinfectant is widely used in privately-owned water supplies or in those serving small communities. In many such cases, the water originates in springs or wells and is already substantially pure. Silver ion treatment is either used to "polish" such waters or to maintain their existing quality during transport or storage. The technology is also used for treatment of water supplies in breweries or foodstuff plants (including in manufacture of pre-packed ice cubes).

Application of silver treatment of water, for example in ships' tanks or on oil rigs has made the characteristic brackish taste and odour of such water, a thing of the past, as it has the risk of stomach infections, especially when drinking water is stored for too long in warm conditions. More recently NASA employed silver for bacteriological treatment of potable water on board the Space Shuttle (111).

As domestic consumers become more demanding of the highest water quality, active charcoal filters and water softeners are increasingly found in private households. After an initial enthusiasm for such appliances, it became recognised that they provided an ideal environment for bacterial growth, especially where the water they contained remained static for days or weeks. Water then drawn off could then be substantially contaminated. The only practicable solution to date, has been incorporation of silver in the active charcoal or incorporating 3–5% of silver-dosed ion exchange resin when charging up the water softener. The use of such silver-dosed ion exchange resin was recommended by the SVGW (Swiss Association of Water & Gas Professionals) (112).

One well-known proprietary process for silver-dosing of water by anodic dissolution is the so-called Elektro-Katadyn® process developed by Krause (109). In a related technology, also developed by Krause, the Katadyn® process, water is passed over or through a large surface area porous structure such as a ceramic or expanded metal, which is then silver coated or impregnated. The silver thereby "corrodes", releasing its ions into solution. (the patents describing this and similar systems are included in the CD-ROM database)

In the UK, Messrs Tarn-Pure provide electrolytic cells in which anodes of a silver-copper alloy dissolve under the influence of a controlled electric current to provide the desired concentration of the ions of the two metals. In trials, a million-fold reduction of bacterial concentration was recorded after 15 minutes, and after 1 hour, the concentration fell to 1000 cells/litre (110).

These systems either employ identical anodes and cathode (of silver-copper alloy) with current reversal, which means that any scale forming on the cathode, is removed when that cathode becomes the anode. In an alternative mode, a pair of silver-copper anodes are positioned either side of a platinised titanium anode. A sketch of such a cell is included in the CD-ROM.

1.13.1 Scale Formation

Wherever electrolysis electrodes are immersed in naturally-occurring waters (river water or seawater), scale is prone to form at the cathode. The reason for this is that hydrogen evolution is taking place at the cathode:

Water dissociates:	$H_2O = OH^- + H^+$
Protons are discharged at the cathode	$2H^+ + 2e^- = H_2$

Removal of these protons results in formation of a near-electrode layer which is alkaline. As a result, insoluble species such as calcium or magnesium carbonates or hydroxides form and adhere to the electrode surface. In so doing, they partially obscure the electrode surface, such that – assuming total current is constant – the current *density* increases and often results in burning or other damage to the electrode surface.

One German manufacturer is Messrs Umwelttechnologie GmbH, based in Gomaringen. Their products fall mainly into two categories:

1) static systems (e.g shower hoses)
2) recirculatory systems (e.g cooling loops)

In the former case, a thick silver wire is wound with a gold-plated steel wire. As a result of the galvanic couple formed between the gold and silver, silver is anodically dissolved and sterilises static water in the shower hose. In the latter systems, an electrolysis cell is inserted in the loop in a by-pass mode. The cell incorporates twin anodes, one of silver, one of copper. Anodic dissolution of the copper provides an algicidal action, that of the silver a bactericidal action. The electrolytic cell is proportioned according to the flow rates. The anodically dissolved silver can be recovered by cathodic electrodeposition, thereby allowing a steady-state concentration to be maintained.

Degussa offer a silver-loaded active charcoal under the trade name "Hydraffin Ag" (113) where the metal is present in finely-divided form, thereby providing a relatively rapid dissolution of the metal to its ionised form. Depending on the characteristics of the water (pH, salts present), between 10 and 500μg would dissolve into the water and 1 litre of Hydraffin Ag can thus be used to treat from 600 to 30,000 litres of water. The material is extensively used wherever a danger exists of contamination of the active charcoal by cell growth. Drinking water, especially when chlorinated, is virtually bacteria-free, and as long as such water is flowing, there is minimal danger of bacterial growth in the charcoal. Under static conditions, contaminating species can enter from the outflow side, and then proliferate rapidly on the charcoal. Silver-loaded charcoal effectively prevents this. A similar product is sold for domestic use as Brita Filter cartridges.

A more recent arrival on the market are so-called Silvana silver membranes, with pore sizes of 0.2 to 5μm. Their pore size distribution is precisely controlled and with their large metal surface area, are ideal for water treatment, especially using electrolytic techniques. Given their very high porosity, high flow-through rates can be achieved (87).

Purification of drinking water for travellers or for use after civil catastrophes, poses the problem that sufficient silver ions must be present to kill whatever burden of cell contamination is present. The problem is aggravated in that the degree of contamination is rarely known to the user and erroneous estimates can thus be disastrous. A recent example of this occurred when the British Army re-took the Falklands, after invasion by Argentina. The British soldiers were equipped with silver-based water purification tablets. Unknown to them, the very high population of grazing animals (mainly sheep) imposed an unexpectedly high bacterial burden on the water, and the silver ions were completely overwhelmed, with the result that the treated water was not in fact sterile. As a result, many soldiers succumbed to diahorrhea and were unfit for duty. This story might be used to make the following point. Whatever means of disinfectation is used, the disinfectant species should always be present in excess, even if only slightly. Though there are numerous analytical techniques for chlorine, it is certainly true that a slight smell of free chlorine in drinking water offers reassurance to the user.

Somewhat analogous to the Falklands tale, the point should be made that many potable water sources contain chloride and some also contain hydrogen sulphide or soluble sulphides. In these cases too, free silver ions will be scavenged to form insoluble silver compounds, and there may then be too low a concentration of these ions to offer a bactericidal action, especially when the presence of such chlorides or sulphides is not suspected. We thus return to the importance of the solubility product, as discussed on page 17 where the concentration of the cationic species (mainly silver, rarely copper) is determined by the nature of the anionic species present, and the solubility of the least soluble of compounds AgX, where X might be a chloride or a sulphide. Most commonly, this will be the chloride from which it follows that the oligodynamic action of silver or copper can be limited by anions forming a sparingly soluble metal salt.

It should be noted here that – in contrast to silver – all commonly found copper salts are reasonably, if not highly, soluble. It is not clear whether the full significance of this has been appreciated, namely that oligodynamically-active copper ions are much less likely to be scavenged from water by commonly found anions, than their silver counterparts. The oxides and sulphides of copper (which is normally divalent, but also occurs in the monovalent state) are the main exceptions to this.

In practical terms, silver as a bactericide has limited utility where high chloride concentrations prevail. As a rule of thumb, a limit of 500mg/l chloride ion is suggested, with virtually complete absence of sulphides and iron concentration < 2mg/l (113).

Such problems have resulted in water purification equipment for travellers being offered which either does not use silver or has dual action based on silver and, for example micro-filtration.

1.14 SILVER-COPPER WATER TREATMENT – SCIENTIFIC & TECHNOLOGICAL SUMMARY

Anyone with more than a casual interest in this, should ask themselves the following questions (for these, and the answers, the writer is greatly indebted to John Hayes of Tarn-Pure Ltd).

Q. Will silver and/or copper ions act as biocides to the species in question?

A. The text, together with the database and other material on the CD-ROM, which can be searched, name most types of bacteria, viruses etc, which have been studied.

Q. What evidence is there of resistance to metal ions developing?

A. Copper-tolerant biofilms have been reported (see CD-ROM database)

Q. Is the use of copper in conjunction with silver simply a means of cost-saving?

A. No, copper with silver can be a more effective biocide than silver alone, at the same total molar concentration. There is a synergy.

Q. Can copper or silver ions be scavenged by other metals in a water system?

A. In principle, yes. Exposed steel or iron are rarely encountered in water systems and could remove copper or silver ions by displacement. Iron hydroxides also. Zinc (hot-dip galvanised) surfaces appear not to present a problem. Some anions (as discussed above) can remove silver ions by forming insoluble species.

Q. Are biofilms formed on silver-copper-electrodes?

A. No, since these are slowly dissolving. They can certainly form on copper surface not acting as electrodes (see CD-ROM database)

Q. Are copper and silver ions released in the proportion they occur in the alloy of which the electrode is made?

A. No. Copper is preferentially released

1.15 MECHANISMS, THEORIES, PROBLEMS

In most applications of oligodynamics, such as water treatment, while a metal or other solid phase may be present, there is an aqueous environment and the general consensus is that it is the metal ions which are the bactericidal or fungicidal agents. True, there are those, such as Batarseh (CD-ROM database) who argue that silver chelates rather than silver ions are the more potent species. But by and large, such questions lie in the field of the biochemist or the microbiologist and would appear, at least to the layman, to be fairly straightforward in scientific terms.

1.15.1 Oligodynamic Effects on Surfaces

Only relatively recently have studies been carried out of the bactericidal action of metal surfaces in air. Keevil and co-workers have been at the forefront of such studies, albeit mainly on copper and copper alloy surfaces. He also cites Robine et al. who have carried out similar studies. (181) Japanese manufacturers have led the way in developing stainless steels incorporating silver, as described both in the text and in numerous patents in the CD-ROM database.

1.15.2 Problems in Study of Copper Alloys as Oligodynamic Surfaces

Workers such as Keevil face a near-impossible task in studying the oligodynamic action of copper-base alloys in atmospheric environment; Such alloys are, without exception, prone to the following actions in atmosphere:

(i) Tarnishing or patination
Copper and the alloying metal(s) react with oxygen to form oxides, with carbon dioxide to form carbonates and with sulphur-containing species to form sulphides. The chemistry of naturally-formed patinas on copper has been studied and is well understood, with simple inorganic salts being formed initially, then being converted over a period of years to more complex double salts. For alloys such as brass, the chemistry is far more complex.

(ii) Dezincification
All copper alloys are prone to loss, by selective dissolution, of the less-noble component(s). Thus brass, under a wide range of conditions, will suffer from selective dissolution of zinc. Bronzes, to a much lesser extent, may lose tin. Loss of the less noble metal can leave a highly porous and usually mechanically friable outer layer of copper or copper-rich material. In many cases, this is then removed by mechanical action and so a general loss of metal occurs.

(iii) Chemo-Mechanical Wear
Everyone will have seen a brass plate or similar feature, where the lettering or decoration has been visibly worn away by assiduous regular polishing. This is mechanical wear and it will be seen that the surface remains smooth, highly reflective and homogeneous - this is what polishing seeks to achieve. If selective dissolution of zinc does occur, the mechanical action of the polishing immediately removes the excess copper. There is no darkening of the surface, nor is the characteristic pink tinge of fresh pure copper seen.

Contrast this with close examination of a well-used copper or (more usually) brass door handle. As installed, this would have been highly-polished, bright, and homogeneous. But after a year or so of regular use, the surface, while remaining smooth, will no longer appear homogeneous. Indeed metallurgists delight in noting that, after extensive use, the underlying grain structure of the metal is revealed. The repeated application of the human hand has brought about an action very similar to metallographic etching. A few microns of metal have been progressively removed by a combined mechanical and chemical action, in the latter case, due to natural moisture such as perspiration present on the hands. Simple organic acids are also present, some of which can form complexes with the copper and zinc present.

In summary, any experimental study seeking to replicate oligodynamic mechanisms on door handles or push-plates or the like, faces formidable problems when attempting to replicate "real life" conditions. It might be noted here that while most well-known metals have shown at least some oligodynamic activity (see the various series shown on page 25), zinc figures regularly in the literature and it might well be that brass, exposed in actual service to mechanical wear which laboratory tests do not usually replicate, might be more oligodynamically active than such laboratory tests suggest.

1.15.3 Corrosion of Copper & Silver and their Alloys

Corrosion is a vast subject, with hundreds of textbooks and tens of thousands of papers and even if one is to limit oneself to the two metals, quite impossible to cover here. However one of the greatest conceptual breakthroughs in the subject was due to Marcel Pourbaix (186) who expressed his ideas in so-called Pourbaix diagrams (now more prosaically known as E_H – pH diagrams). Derived from fundamental thermodynamic calculations, each representing a particular chemical or electrochemical equation, these diagrams show the behaviour of a metal immersed in an aqueous solution. They indicate the state of the metal surface – metallic or oxidised (passivated) and what reactions such as hydrogen or oxygen evolution are possible. The diagrams show all this as a function of two variables – solution pH and the potential of the metal. What determines the potential of a given metal under any particular set of circumstances, is another matter. It may be that an external current source is applied, or that the metal forms one half of a galvanic couple, or that some other electrochemical reaction is taking place and driving the potential of the metal. However the main point is that both pH and metal potential can readily be measured and thus, by doing so, one can predict or determine the state of that metal.

Pourbaix diagrams for copper and silver are shown below: (180)

Figure 1.6, like all Pourbaix diagrams, plots pH versus potential and shows how copper, immersed in aqueous solution, at 25°C will either

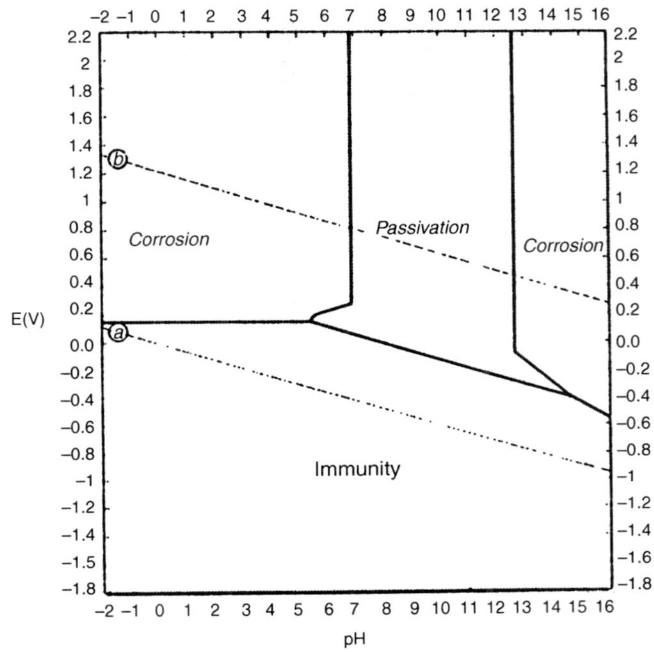

Figure 1.6 Simplified Pourbaix Diagram for Copper in Aqueous Media.

be "immune", that is not corroding, passivated by forming a protective oxide on its surface, or it will be actively corroding – releasing copper ions into solution. It is noted that it can corrode in two separate zones (and conditions), the first being corrosion of bare metal, the second (so-called trans-passive), at high potentials where the presence of oxide or other species on the surface is no longer sufficient to protect the metal.

The two dashed lines (a) and (b) represent the theoretical potentials at which hydrogen or oxygen evolution respectively, can take place. In certain circumstances, these reactions may form part of an overall corrosion process.

It is important to note that the diagrams above are a greatly simplified form of Pourbaix diagram. In their more advanced forms, these show different oxides or oxidation states, and show how, for example, the borderline between corrosion and immunity depends on the concentration of the metal ion (in this case copper) in solution. These diagrams remain invaluable in that, because they plot two readily measurable parameters, one can readily establish what the surface condition of a metal is likely to be, and thus how it might be expected to behave.

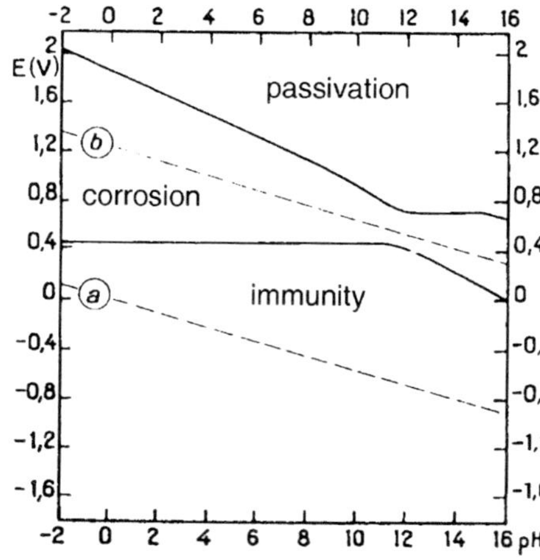

Figure 1.7 Pourbaix Diagram for Silver.

1.16 ATMOSPHERIC CORROSION OF COPPER ALLOYS

If the oligodynamic action of copper and alloys as door handles etc is due to release of copper ions, this takes us to a specialised branch of corrosion science known as atmospheric corrosion. This is based on exactly the same principles as "wet corrosion" except that it relates to a situation where only a thin film of moisture is present on the metal surface. Now, factors such as relative humidity and dew-point come into play, in establishing how much moisture will be present. In some circumstances, where corrosion involves, as a counter-reaction, the reduction of oxygen from air, the corrosion rate can be more rapid than under "immersed" conditions, since the oxygen can diffuse more rapidly to the metal surface through a thin moisture film than it could through a larger mass of water.

There have been many studies of the atmospheric corrosion of copper and alloys. Most of these consist of exposing weighed coupons of the metal or alloys in various outdoor sites, inclined at an angle of 30° or so to the vertical. After several months, the panels are removed and re-weighed and weight-loss is recorded, usually after removing of corrosion products. Data of this kind is reported by Leidheiser in his book, by Rozenfeld in his volume and in an ASTM STP volume edited by Dean & Rhea (see CD-ROM database) to name but a few sources. But how close an analogy does such work provide with the day-to-day environment of a push-plate or door-knob?

Leidheiser's book also includes a short chapter on the effect of micro-organisms on corrosion of copper and its alloys, which he shows can considerably accelerate the process. Keevil too has studied the increased rate of pitting of copper in presence of biofilms (CD-ROM Database).

However there is a further element of possible importance. The whole process of corrosion in an aqueous or pseudo-aqueous environment (as one might describe a door knob in regular use) is governed by the free energy change between the metal in its initial metallic state, and the corrosion product it forms. Thus we can write:

$$M =====> M^{+} + e^{-} \text{ and } \Delta G = -nFE$$

where ΔG is the change in Gibbs Free Energy of the overall process, n the number of electrons transferred (in this case 1), F the Faraday constant and E, in effect, the corrosion potential. We might construct such an equation for copper, (in which case $n = 2$). The unspoken assumption is that the reaction product, Cu^{++}, is a hydrated copper ion. But supposing that rather than being a hydrated metal ion, the product is a copper complex ion of a species found in perspiration, such as a lactate or a chloride ion-complex ? In such a case, the corrosion rate of the copper (or alloy) could change substantially. There is a large body of literature describing corrosion caused or enhanced by finger-marking and perspiration contacts. The principle of shifting corrosion potentials by introduction of complexants, is widely made use of in electroplating (the reverse of corrosion), where introduction of metal ion complexing agents totally changes the energetics of a given system. Thus lactic acid, found in sweat, is widely used in electroless nickel plating.

Alternatively, one might take a lateral leap to the supposedly therapeutic benefits of wearing copper bracelets. Leaving such benefits aside, it is certainly the case that – more with some wearers than others – the copper corrodes and stains the contiguous skin and it seems likely there is a transdermal uptake of the copper.

That micro-organisms can be "electro-active" is not in doubt, as work of Keevil and Leidheiser cited above, makes clear. Just as atmospheric corrosion is a subset of corrosion science, so also is biocorrosion. By this is meant that such species can take part in an electrochemical process such as corrosion, and affect the outcome. Brain (184) in unpublished work, carried out EIS (Electrochemical Impedance Spectroscopy) studies on a range of bacteria at stainless steel surfaces and noted that, once killed, the electrochemical response changed from being primarily capacitative to primarily inductive, thereby confirming the electro-activity of live, but not dead bacterial matter.

If, as most believe, the oligodynamic action of copper surfaces involves its corrosion to form copper ions, this raises the whole question of what electron-acceptors are available to allow this. Pourbaix diagrams (q.v) are predicated on hydrogen evolution or oxygen reduction in this role, but it seems clear that at least some micro-organisms might act in the same way. Dead bacterial matter might, in theory, obscure the metal surface, though this is thought unlikely – it is most probably mechanically dislodged.

The work of Keevil and his team is of undoubted value but the observations above suggest we have yet to replicate in the laboratory and to fully understand, the very complex events taking place on a metal surface in a real-life environment. Only in-situ experiments can give a definitive answer, and trials are presently under way at Selly Oak (Birmingham) hospital, the results of which are awaited with much interest.

1.17 KINETICS & MECHANISM

There have been few attempts to elucidate the kinetics and mechanisms of oligodynamic processes. As regards the reaction of bacteria on metallic surfaces, at least in theory, graphs published in various papers by Keevil, and the data in DE 4324994 (reproduced on page 70) might form the grist for such analysis. Where graphs compare the behaviour of various copper alloys, most authors specify their composition in weight %. Possibly comparisons based on atomic % would reveal trends more clearly. However the data is hard to interpret. The patent reports two sets of experiments. In the first, the initial cell count was 3 million, in the second, it was 30,000. The data is a 3 × 6 matrix. Test species were *Staph. Aureus*, *Staph. Epi.* and *Enterococcus*. Cell counts were taken in each case at 1,6 and 24hrs and 6 different test surfaces are reported and cover up to 6 orders of magnitude. The following questions might be asked:

Q.1 Is the kill-rate consistently higher for any one of the three species named above?

A.1 No consistent ranking is found. The data does not indicate any one species to be consistently more vulnerable than another.

Q.2 Can any kill-rate kinetics be adduced?

A.2 Not with confidence. Thus *Staph. Epi* declines from 3 million to 100,000 after 1 hr and this value remains unchaned at 6 and 24hrs (on nickel-coated brass). On powder-coated aluminium, the *Enterococcus* cell count increases from 1 million (6 hrs) to 2 million (24 hrs). In the case of a silver surface, the cell count does not change between 1 hr and 6 hrs for *Staph. aureus* or *Enterococcus* though it does drastically decrease at the 24 hr point

Q.3 Can one rank the oligodynamic activity of the 6 surfaces?

A.3 In broad terms, the copper-tin surface appears to be most active. This is the basis of the claim in the patent. Silver and anodised aluminium (surprisingly!) appear as second-best. Nickel and powder coated aluminium (an organic resin) score worst. Stainless steel, which appears in Keevil's work as almost inert, at least for the first 3 hours, here exhibits a bactericidal activity which is not negligible (*Enterococcus* down from 3 million to 100 after 24 hrs). This might accord with Keevil's observation that after an initial 3 hr induction period, stainless steel too, exhibits an oligodynamic action.

Keevil's data is also difficult to fit to any simple kinetics. In his study of influenza virus (185) he starts with 2 million virus particles (as compared with 3 million in the patent above) and also uses 6 and 24 hour sampling times. In the case of stainless steel, he still finds 500,000 particles after 24 hrs, many orders of magnitude greater than in the patent (given the same cell species was not used). On copper, by contrast, Keevil finds a count of 500, not dissimilar to the figure of 100 (again for different species and on silver).

Keevil et al. have made at least one attempt to fit their bacteriological results to the composition of the various copper alloys studied by them, by plotting time for (essentially) complete kill vs. copper content of the alloys. (179). Fig 1.8 is an idealised graph based on Keevil's data.

In Figure 1.8, the decay curves show (from top to bottom) stainless steel, where no bacterial kill was observed in the first 100 minutes, though there was a log-linear decrease in cell count after that. The next plot is silver nickel (UNS C75200) with 65 wt% Cu, 17wt% Zn, 18 wt% Sn. The next curve is copper-nickel (UNS70600) with 89% Cu and 10% Ni while the lowest curve showing a cell count of log 0 at 60 minutes can be taken to represent pure copper, brass (90% Cu, 10% Zn) and a range of bronzes with copper content between 90 and

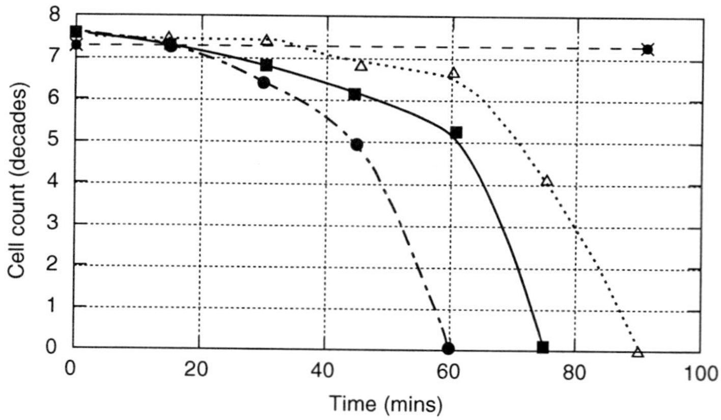

Figure 1.8 Idealised plot showing survival of *Listeria monocytogenes* Scott A on different metal surfaces in air (see text).

97% copper. Keevil's original data not only shows quite clearly, an "induction period" for the kill rate on stainless steel but also shows a plateau or "rest period" of some 20 minutes in length in the case of pure copper and the high copper alloys, during which no significant bacterial kill takes place. Whether this is an experimental artefact or a genuine phenomenon indicative of a change of mechanism, cannot be stated. His data does, however, unequivocally establish a relationship between copper content and bacterial kill rate.

In other work (173) Keevil et al. show what appears to be an induction period for *Listeria* kill on stainless steel, with almost zero kill rate in the first 100 minutes, followed by a log-linear kill of 2 orders of magnitude over the following 3 hours.

1.18 THEORY & MECHANISMS – CONCLUSIONS

It is clear that, in terms of oligodynamic effects on air-exposed surfaces, the science is still at an early stage. Should one seek to study more closely an actual in-service door-knob or push-plate ? Or should one design a laboratory experiment which embodies some or all of the factors noted above ? The foregoing discussion leaves no doubt that complex chemical-mechanical-electrochemical actions are at work. That metal(s) are removed from in-service door handles etc, is not in doubt. However the amounts in question are so small that only the most sophisticated techniques would be able to quantify them. An understanding of the various mechanisms would facilitate the design of metallic surfaces not unlike the silver-loaded stainless steels discussed elsewhere, in which dissimilar metals or alloys were combined to ensure an optimum rate of metal ion release. However in the mean time, the experiments at Selly Oak Hospital should reveal, whatever the mechanism, how efficacious copper-containing surfaces are in control of nosocomial infection.

PART 2

APPLICATIONS OF OLIGODYNAMICS IN MATERIALS

2

Oligodynamic Effects of Silver & Copper on Functional Surfaces in Medical, Food Industry & Related Environments

2.1.1 General Comments and Basic Principles – Contaminated Surfaces

The use of metals such as silver and copper for water treatment appears to be a relatively mature technology; it is not seriously challenged in these applications, by what might appear to be competitive methods. The CD-ROM database includes several abstracts which describe comparative studies of the efficacy of other methods including chlorination, thermal cycling and UV irradiation. Of the last, Hayes (174) notes that not only is the UV radiation blocked by bubbles in solution, leaving unsterilised areas in their shadow, but also – unlike most other methods – it does not normally generate bactericidal species capable of existing for any length of down and able to act downstream of the UV source

How times change! The writer recalls a fellow student, the butt of much laughter and joking from his peers, who invariably opened doors using his handkerchief, or perhaps his coat flap, to avoid direct contact with the metal. Today we know that his "hypochondria" was not misguided, and that all manner of solid surfaces can carry a rich harvest of germs which can survive for many weeks. We are indebted to Kramer & Schwebke (CD-ROM) for a most thorough review of this. They bring together a collation of studies where the effects of time, temperature and humidity on the persistence of a wide range of micro-organisms on solid surfaces, is considered. Their work is strongly recommended to anyone with an interest in a particular type of micro-organism or pathogen. However the effect of the actual surface material is not, by and large, made explicit in their work. Possibly there are interesting conclusions to be "mined" from their data – a task for some future worker?

The whole concept of "bactericidal surfaces", that is – surfaces which incorporate a means of actively killing micro-organisms – is relatively new. Ten years ago, the term itself, and related matters such as bactericidal paints or other coatings, were virtually unknown. Today, scarcely a week passes without some news of a novel coating or surface, or a new application for such materials. This terminology may be relatively new, though history does in fact reveal examples of bactericidal or similarly active surfaces. Of these, none is better known than the use of copper sheathing of wooden ships' hulls. Partly this served as a mechanical barrier to ingress of wood-boring marine creatures. But equally, use of the metal was found to inhibit fouling of the hull by seaweed and other vegetable growths, as well as discouraging molluscs. Given what we now know, it is clear this is an early example of the application of oligodynamic effects. A somewhat less immediate example relates to the use of wooden chopping boards, found in virtually every kitchen. A few decades ago, food hygiene "experts" demanded the removal of these, certainly from commercial premises subject to legal inspections. They decreed that such "unhygienic" items be replaced with those made of plastic, such as polypropylene which were non-porous and thus less able to harbour micro-organisms. Over time, however, further study suggested that the natural oils, possibly tannins, present in hardwood chopping boards were themselves bactericidal. The pressure to replace such wooden items with a plastic counterpart appears in many cases to have been quietly abandoned. It is thus clear that, whether it was explicitly recognised or not, the concept of a bactericidal or bioactive surface, is thousands of years old, as examples in Chapter 1 indicate.

As will be seen, there are many different ways in which the oligodynamic action of a metal can be harnessed to create a material with an "active" surface. In the case of copper, the price of the metal (though it has greatly increased in recent years) is still low enough to allow the metal to be used in massive or solid form. In the case of silver, this can also be the case but more rarely so, outside of medicine where, for example, catheters or tracheotomy tubes can be made of, or incorporate, massive silver. In today's climate, materials tend to be designed according to two quite separate philosophies. On the one hand, is the philosophy (notably in the case of precious metal catalysts) where the aim is to use as little of the material as possible. Two ways in which this can be achieved are by applying a material as a coating, and making that coating as thin as possible. In other circumstances, the aim will be to maximise the surface area:weight ratio. Here, the two classic approaches are to make the particle size as small as possible, since this improves the surface area:weight ratio. Alternatively, one can seek to "extend" the use of an active material, either by diluting it in some way or (again as with catalysts) using a "support" material which may act in a purely physical manner or may exhibit some form of synergism.

The second philosophy which has come much to the fore in recent times, most notably in the automobile industry, is to use materials in such a form that they can readily be recycled. An obvious area for this approach would be the use of silver in medical appliances as noted above. Sadly, though, security then becomes a problem. One runs a risk that a valuable piece of equipment might be stolen simply for the scrap value of the silver it contains.

2.1.2 Options for Creating Oligodynamic Metal Surfaces

The main options can be summarised as follows:

1. Use of massive metal; Examples: copper or copper alloy door handles, silver catheters.
2. Continuous metal coatings. Examples: cladding, plating, coating
3. Dispersed (particulate) metal. Examples: silver-loaded paints, plastics, ceramics. Stainless steel containing finely-divided silver particles

The descriptions of these various methods that follow, are deliberately conceptual rather than being detailed and specific methods. Their practical execution can be found in numerous metal finishing textbooks (see, for example, www.finishingpublications.com) and browse the CD-ROM database

2.1.3 Silver Coating or Cladding

Commenting on the above, there is little to be said as to 1. It is assumed that the items will be fabricated by the usual metalworking techniques. In terms of coating, the question arises as to how thick to make a coating, which will be subject to varying degrees of mechanical wear as well as the corrosion which releases the metal ions required to act oligodynamically. Cladding is a process by which a sheet of copper or silver is bonded to a metal substrate. The thickness of such coatings is determined by limiting factors at either extreme. For the thinnest coatings, it may be more cost-effective to apply the metal by other means (e.g electroplating). Additionally, the cladding metal (in sheet form) must possess sufficient mechanical strength to allow it to be handled. Apart from gilding (using gold foil) a lower limit for cladding would be of order 100μm. At the other extreme, the sheet of cladding metal must retain sufficient ductility for the bonding process (usually by rolling) to be viable. An upper thickness limit of order 1 to 2mm is a broad guideline value.

For thinner coatings, electroplating is the default method. The electroplating of silver or copper are both carried out on a large scale in every industrialised country and are used to deposit coatings from typically 2μm to 20μm thick. However both thinner coatings (usually referred to as "flash coatings") and thicker coatings can be formed without difficulty. In the latter case, the coating may be so thick that it no longer requires a substrate to support it, that is, the coating can function in free-standing mode. This is usually referred to as electroforming and electroformed copper or silver functional or decorative items have been made for around a century, typically with thickness of order 1mm.

As candidate coatings for oligodynamic use, two questions will be asked of electroplated (and other) coatings. The first question is – will the thickness provide a sufficient reservoir of metal to fuel the oligodynamic action? Thickness can be converted to weight using the simple relationship:

$$\text{Mass} = \text{Volume} \times \text{density}$$

(note that the density of electroplated metals is always slightly less than for their massive form, typically 90–95% of the bulk value, as a result of porosity in the deposited film). The rate at which the metal will be released is far more difficult to calculate, depending as it does on the application in question and the degree of use to which it is subjected.

However some useful analogous cases can be drawn on. Silver-plated cutlery and utensils have been sold for over a century and there is extensive experience as to the longevity of such coatings. In practice, silver-plated cutlery is available in a range of qualities, distinguished mainly by the thickness of the plated metal which ranges from 10µm for the cheapest quality, to 50µm or so, which in the past was "guaranteed for life". In many countries, the thickness of such coatings is governed by national Standards. Certainly in the case of thinner coatings, wear is routinely evident, most commonly on the underside of the bowl of the spoon or the lowest point of a fork, where it rests on the table surface. Turning to copper plating, in recent years and only after extensive trials, coins made of solid copper (with minor alloying components) have been replaced by copper-plated steel coins. Though undoubtedly subject to considerable wear of various forms, one very rarely observes any "show-through" of the underlying steel. Finally, mention might be made of antique tableware known as "Sheffield Plate". This is no longer manufactured, as far as the writer knows. It was made by mechanically cladding copper substrates with silver sheet of thickness around 100µm. Examples can be found in many antique dealers. In many cases, mainly due to over-enthusiastic polishing, the silver cladding has worn through and the underlying copper is visible.

These examples are provided to indicate that, in many cases, silver coatings and probably those of copper too, will sooner or later wear away, and in designing oligodynamic systems, this should be taken into account.

In certain circumstances, so-called "electroless" copper or silver coatings can be formed. An object (which can be metallic, plastic or ceramic) is immersed in the electroless plating bath, whereon a coating forms spontaneously over its entire surface. Such processes are inherently more expensive than electroplating, but in certain circumstances, may be preferable or indeed the only option. Examples of the latter are when the substrate is not electrically conductive, or when the substrate has a very complex shape and there is a need to coat all its surfaces with a uniformly thick film.

There are other coating processes, but in the present context, it is not easy to see where they might be used. So-called "immersion plating" or "displacement plating", also known as cementation, is based on the atom-by-atom replacement of a less noble metal with one more noble. Thus, a copper coin immersed in silver nitrate solution will become "silvered", as copper atoms go into solution, and are replaced by silver ions, reduced to form silver atoms. Such coatings are usually of poor quality and are inevitably thin, since once one or two layers of the copper have been replaced with silver, the underlying copper is then obscured and no longer available to be replaced by silver. This process can also take place as an undesired one. If a water system is being dosed with silver and/or copper ions, these could in theory at least, deposit onto any exposed steel surfaces, or those with hot-dip galvanised (zinc) coatings. Indeed, one could conceive of copper ions behaving in this way to form metallic copper particles on the surface, with these, in turn, forming sites where silver ions were deposited to form metallic silver, by exchange with the metallic copper. Fig. 2.1 below shows an idealised process, silver ions are depositing from solution to form particles of metallic silver, while copper is being ionised (corroded), releasing copper ions into solution on the basis of charge neutrality, that is, two monovalent silver ions deposited for every divalent copper ion formed.

Another family of coating processes are those employing vacuum techniques. PVD (Physical Vapour Deposition) or CVD (Chemical Vapour Deposition) are widely used in the semiconductor industry, for example and for metallising plastic bottles, mainly for aesthetic purposes, as with perfume containers. Though usually operated in batch mode, they can be used continuously, for example to coat reels of plastic film. These methods are not usually advocated for coating of more than a few micron thickness and it is not apparent in what circumstances they

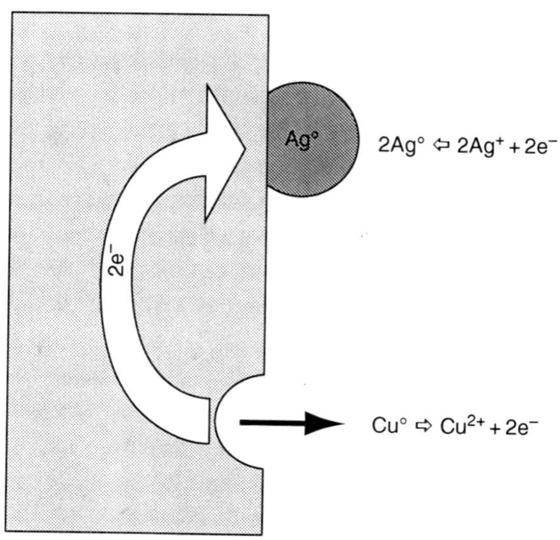

Figure 2.1 Idealised cementation mechanism.

might be relevant here, other than to provide oligodynamic action for "throwaway" or disposable items.

Finally, a mention of a widely-used coating process, certainly used for copper, but not, as far as the writer is aware, for silver. This is flame-spraying or plasma spraying. The metal, in powder or wire form, is passed through a high-temperature flame such as oxy-acetylene, where it is melted, and sprayed from a spray gun onto the substrate. Coatings of almost any thickness can be formed. Usually these are somewhat porous, but as a means of applying significant quantities of these metals, the technique should not be overlooked.

Most of these coatings can be applied with a range of surface roughness values, from near-mirror finish to the very rough (CLA values of 50μm or more). The surface roughness of the substrate on which a coating is applied, will also affect the final roughness of any coating applied to it. Leaving aesthetics aside, a rough surface finish has a larger true surface area (geometric surface area x roughness factor = true surface area). There is a near direct proportionality between true surface area and the rate of metal ion release, implying that smaller apparent active surface areas can be used. Against this, rougher surfaces are more difficult to clean (and can thus harbour more undesirable organisms) and can provide superior anchorage for biofilms, which can be equally undesirable.

2.1.4 Which Coating Method?

We see, then, that a range of processes are available for the coating of almost any substrate with copper, silver or some other alloy. Which is the most appropriate? The answer depends mainly on the thickness of the oligodynamic metal desired, and that, in turn, is mainly a question of how large a "reservoir" of the active metal is called for. In the case of one-shot water treatment kits, very little metal is required since only a few litres of water will be treated. For permanent installations, designed perhaps to operate for 12 months between inspections, use of massive metal rather than copper or silver coating might be a better solution. In general terms, vacuum or gas-phase coating techniques such as PVD or CVD, are preferred for thin (sub 1μm or so) coatings. Electroless plating (mostly used for sub-10μm or so thicknesses) is more expensive than electroplating. Hot metal spraying is probably the least expensive.

The foregoing comments have mainly been directed to coatings applied to a broadly planar surface. However the metallising of open-pored foam structures, such as porous plastics, is well-known, and in this case electroless plating is most commonly used. The coating thickness here will depend on the pore diameter, too great a thickness and the pores will become more or less completely blocked, thus defeating the object of the exercise, which is to create a high surface area through which gases or liquids can flow. In some cases, after metallising a polymer foam, the material is heated in air, pyrolysing the polymer and leaving a purely metallic foam behind.

2.1.5 Silver in Dispersed Form

Metal-loaded paints, plastics, ceramics, graphites or other materials are nothing new. They fall under the general heading of composite materials. In recent years, there has been a trend to move from micron-sized particles to much smaller ones, of nanometer dimensions, known as nanoparticles. In many cases, this change has improved the performance (however defined) of such products.

In considering such composite materials, the default assumption is that the matrix material is inert, and that the metal particles contained within the matrix are immobile. The implications of this are that once those metal particles at or very close to the surface are released, the remainder, more deeply embedded in the matrix, are inaccessible and thus useless. There are several solutions to this. The first is creation of so-called “compositionally graded” coatings where the distribution of the particles in the matrix is non-uniform. In the present case, one would seek to bring about a much greater concentration of silver or copper particles close to the outer surface.

A second approach is one in which the matrix is not totally inert, and by itself wearing away or dissolving away, it exposes fresh surfaces and metal particles whch were originally at some depth from the outer surface. A third possibility would be to create a composite in which the metal particles did possess a degree of mobility. Then, as the outermost particles were removed by wear or ionisation, those further from the surface would redistribute themselves in response to the concentration gradient created as the outermost particles were removed. Browsing the CD-ROM database reveals the breadth of such approaches.

2.1.6 Metals in Non-Metallic Form

As previous chapters have made clear, the metals themselves, such as silver or copper, are inactive – they are no more than potential sources of metal ions. It thus follows that compounds of silver or copper are, at very least, to be considered as candidate materials for incorporation into oligodynamically-active surfaces. The degree of solubility in water is critical here and while many simple silver compounds are highly insoluble in water, this is not the case for copper compounds.

It should be noted that silver occurs naturally either as the metal itself or as an oxide, a salt or a sulphide. Without exception, in its natural form, silver is monovalent. It is not always appreciated that a divalent form of silver exists, but that it is formed only under the action of the strongest oxidising agents, or by electrochemical oxidation. The Ag^{++} ion is highly reactive and will oxidise almost every organic product. This extreme reactivity was studied by Fleischmann et al in the 1970's and was then commercially exploited using a technology where, after chemically oxidising the species to be destroyed, the silver (II) ions are reduced to silver (I), then returned to an electrochemical cell where they are re-oxidised to Ag(II) once more. In recent years, a number of bactericidal compounds based on silver (II) complexes have been reported and patented. Their bactericidal action is not yet clear, and so reactive is Ag(II) that it may well react with whatever species it first encounters, and then be reduced to the monovalent state. In the context of this book, its main application would appear to be that of an additive to water for its treatment. The database includes abstracts on divalent, and even trivalent silver (use these as search terms in the CD-ROM database)

2.1.7 Anodic Oxidising (Anodising)

It is well known that certain metals, such as aluminium and titanium, as well as others often referred to as "valve metals" (on account of their rectifying properties) can form thick oxide coatings when they are positively charged in a suitable electrolyte solution. The process is known as "anodising" and because this oxide coating is both hard and corrosion-resistant, is very widely used to protect fabricated aluminium, when used, for example, in architectural applications as window-frames. The structure of the oxide coating formed on aluminium, is unusual – it is highly-ordered and consists of a series of "cells", each with a pore at its centre. Over the years, a range of applications have developed based on the infiltration, of these pores, with a material which enhances or extends the properties of these oxide coatings. Most common, is introduction of a metal, such as tin, which creates the characteristic "bronze" coloration often found in architectural applications. There is no reason why, should it be so desired, silver or copper (or both) should not be introduced in the same way. Nor is this oxide infiltration technology limited simply to metals. A range of other materials including fluoropolymers and fluorescent species can also be introduced, and thus, it can be assumed, sparingly-soluble silver salts.

A variant of the conventional anodising process, is the so-called plasma anodising, also known as PEO (Plasma Electrolytic Oxidation) or MAO. (Micro-arc Anodising). In this case, much higher voltages are used and the resulting coating is thicker, and often glassy in nature. More importantly, the chemical composition of these coatings is no longer overwhelmingly the metal oxide. Instead, such coatings can include silicates, phosphates, borates and a range of metal salts. PEO coatings are being used in a wide range of applications from engines used in Motor Sport, to Leisure goods such as fishing tackle. However there is a growing interest, expressed in both publications and Patents, in PEO coatings for biomedical implants, notably those where the coating is a phosphate or a hydroxyapatite, ideal for the osseointegrative process necessary for a successful medical operation. Perhaps unsurprisingly, the incorporation of silver in such coatings has been suggested, as in US 2005221259 "Dental or Orthopaedic Implant" (2005, to Plasma Coatings Ltd) It should be noted, however, that the wording of this patent is misleading - it refers repeatedly to "silver particles" whereas it is clear that, as produced, the silver would be in the form of a metal salt, most probably dissolved in the glassy phase of the coating.

2.1.8 Controlled Release

The term "controlled release", also known as "sustained release", has been used in earlier chapters, and the science and technology whereby an active substance is provided in such a form that it is released into its environment over a predetermined period, which might be days, weeks or months, is now established in it own right (175). It is clearly relevant in the terms of this book. There are various means by which controlled release is achieved. One is to encapsulate the active species using a membrane, through which it can diffuse only slowly. Another approach, referred to above, is to embed the active species in a matrix which itself slowly dissolves. Dentistry is one field where the latter approach has been used, where silicate dental cements contained fluoride ions, which they slowly released. These were then replaced by so-called glass ionomer cements, which acted in the same way, though with otherwise superior chemical-mechanical properties.

2.1.9 Alloys

It may have been noted that few scientific studies have been previously cited in which metal alloys, rather than single metals, were studied. This in spite of the fact that, as noted, some commercial water treatment systems use anodes of copper-silver alloy. Only more recently have such studies been carried out, and will be considered below. Alloys should be considered under two headings. Firstly, in terms of oligodynamic activity, how do mixtures of silver and copper ions (whatever their source) compare with that of each metal alone? Is there any synergy at work, where for example the one type of metal ion attacks cellular organisms in a different way or at a different location, from the other? Or to what extent can the less expensive copper ions accomplish some of the bactericidal action, reducing the amount of silver required? Such questions lie in the realm of microbiology and related disciplines.

The second question, which falls in the realm of the metallurgist or corrosionist, considers the corrosion and dissolution of the alloy. If the alloy has two components A and B, will these dissolve at the same rate (e.g atom-for-atom)? Or will there be some form of selective dissolution where A dissolves much faster than B? An example of the latter is the well-known dezincification of brass. This alloy contains copper and zinc (sometimes with other metals). In certain situations, the zinc dissolves much faster than the copper, which is left as a friable sponge, full of holes formerly occupied by the zinc atoms. This fragile copper residue may then crumble away, exposing fresh brass, allowing the cycle to repeat itself. The actual behaviour of such alloys will depend in part on their composition, including the number of different phases present, but also on the conditions - pH, temperature and the potential at which the alloy finds itself, in a manner analogous to that shown in the Pourbaix diagrams in Chapter 1.

2.1.10 The Corrosion Rate of Metals

As discussed in Chapter 1, where silver or copper are present as compounds of these metals, their concentration in solution is determined primarily by the solubility product of the species in question. Where the metal is present in its metallic, i.e uncombined, state, the metal ions which form the basis of the oligodynamic process are formed, loosely speaking, by corrosion of the metal.

We begin by setting down the reaction:

$$Ag \Longrightarrow Ag^{+} + e^{-} \quad (2.1a)$$

Or:

$$Cu \Longrightarrow Cu^{++} + 2e^{-} \quad (2.1b)$$

These are so-called half-cell reactions. They can only take place in circumstances where there is some means available of accepting the electron(s) which is produced by the reaction. In broad terms, there are two mechanisms by which this can occur. These are:

A. "Driven Systems"

An external source of electricity (usually DC) is connected, the +ve lead to the silver or copper, the −ve lead to a so-called counter-electrode, which might be made of steel, nickel, stainless steel etc. An electricity supply can be considered as a source and "sink" for electrons, and thus able to drive a reaction such as those shown above. A second reaction is also involved, where electrons are taken up at the cathode, such as:

$$2H^{+} + 2e^{-} \Longrightarrow H_2 \text{ (evolved as a gas)}$$

The silver or copper is then said to be subject to "anodic dissolution". This is the basis on which the more sophisticated water-treatment installations manufactured by TarnPure, and other manufacturers, operates. Thanks to Faraday's Law, the amount of metal released in its ionised form, into solution, can be precisely calculated. Thus 1 Faraday (96,500 Coulombs or Amp-seconds) will release 108 gm of silver (108 being the atomic weight of silver) or 32 gms of copper (64 being the atomic weight of copper, but 2 electrons are required per atom in this case).

B. Freely Corroding Systems

An external supply of electricity is not necessary to allow reaction (2.1 a or b), above, to take place. Instead, some other "half-cell" reaction may provide a means of taking up the electrons formed when a metal ionises in solution. Two such reactions are the reduction of oxygen (as air dissolved in water) or the evolution of hydrogen, as shown above. Inspection of a Pourbaix diagram (see pp. 43–44) allows at least a qualitative prediction. These diagrams show two parallel hatched lines. The upper one denotes the formation (gas evolution) or reduction of oxygen. The lower one shows the evolution (reduction) of hydrogen gas from its ions. For either of these reactions to function as a counter-reaction, there must be a nett release of energy. In broad terms, this is the case when the metal ionising reaction lies below (at a more negative potential) than the counter reaction. Silver being a relatively noble metal, this is never the case for hydrogen evolution. For copper, as the diagram shows, under some conditions it can be. Oxygen reduction however, always fulfils this condition. A freely-corroding metal might be compared to a leaf floating on the surface of a pool of water. It can be "blown" hither or thither, depending on the prevailing conditions. The metal attains what is sometimes known as a "rest potential" or "open circuit potential" (OCV) which is a function of whatever reaction(s) are taking place at the metal surface, in terms of species present in the water. Finally, let it be noted that silver is more noble than copper and that being so, will in most cases corrode – that is, release metal ions – more slowly than copper, other conditions being equal.

Another frequently encountered type of corrosion is the so-called "galvanic couple" or dissimilar metals corrosion. This is, in effect a miniature battery which has been poorly short-circuited, for example by means of a thin film of moisture on the surface. Such a galvanic couple is a perennial problem in engineering construction, but – as previously indicated – it is sometimes deliberately created in oligodynamic systems, since the corrosion is, in fact, a means of slowly releasing the bactericidal metal ions.

Figure 2.2 shows a metal covered with a film of moisture. A is an anodic region, where metal ions (M^{z+}) are released into the moisture film. K is a cathodic region where, for example, oxygen from air might be reduced. One could easily imagine a copper door-knob, covered with a moisture film too thin to be seen by the naked eye. Part of the door-knob might be bright metal, polished perhaps by regular contact with the hand. Other parts of the door-knob surface might be oxidised or sulphide-tarnished, and these dissimilar regions would create a "cell" which would slowly release metal ions. Or the door-knob might be made of brass, where some areas were zinc-rich, others copper-rich and again,

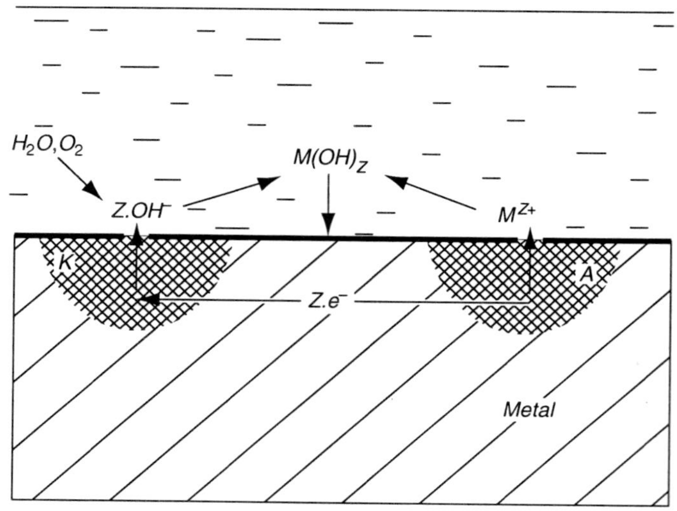

Figure 2.2 Schematic representation of local element corrosion.

these would act as dissimilar metals, capable of forming an electrochemical cell. Described elsewhere in the text, is the formation of a galvanic couple by incorporating local areas of a noble metal such as platinum, at which oxygen reduction proceeds favourably. By determining the size of such local areas, the rate of release of copper or silver ions can be designed into the system.

C. Driven vs. Freely-Corroding Systems

Driven systems, at least in theory, are almost infallible and whatever the temperature or pH of the water, the product of current x time will determine the amount of metal ions released into solution according to Faraday's Law. Freely-corroding systems are much less predictable and variation of temperature, pH, the presence of salts (especially hard water calcium or magnesium salts) and presence of biofilms are but a few of the relevant parameters. As the copper or silver wears away, its rate of corrosion (dissolution) may also decrease. It may be that rather than simply ionising, that some poorly-soluble corrosion products such as carbonates or sulphates, form on the metal surface. These can "shield" further dissolution of the metal. An *a priori* estimate of the rate of metal ion release must therefore be speculative, and the system designer faces the difficult task of over- or under-estimating the metal release rate. The first is wasteful, the second may compromise the functioning of the system.

Perceptive readers may have asked themselves – what of the situation where, rather than using a single metal, an alloy of silver or copper is used? This thorny issue can be sidestepped by using separate anodes of copper and silver (or any other metals). These may be energised each with their own power supply, and though inherently more expensive, is the most reliable approach. Alternatively, separate metal anodes can be used, but fed from a single power source. By varying the relative sizes of copper and silver anodes, the required rates of copper and silver dissolution can be obtained. The distribution of current between the various anodes is too complex to be considered here, other than to note that – weight-for-weight – silver dissolves at roughly the same rate as copper.

Where an alloy, such as copper-silver is used (as is the case for some commercial installations), the situation become complex. It is then a question as to whether uniform or selective dissolution of the two components (e.g silver and copper) is taking place.

2.1.11 Practical Measurements

What simple measurements can be carried out to monitor the behaviour of an oligodynamic metal immersed in water? The easiest of all, is periodic recording of weight loss. Measurements should be made at monthly intervals in the first instance. A weight loss of less than 1% is too small to measure reliably under field conditions, while a change of more than 5% suggests more frequent measurements, perhaps at weekly intervals. The results must be recorded graphically, to give a longer-term view of the process. Matters are complicated where solid corrosion products are formed on the metal surface. These can be removed by suitable mechanical or chemical means, however since these products themselves probably contain the metal (as an insoluble salt), to exclude them from a weight measurement is, in principle, misguided.

Two other measurements can usefully be made, most likely using the same instrument. These are the pH of the water and the potential of the oligodynamic metal, versus a standard reference electrode. A pH meter is in fact no more than a voltmeter with very high input impedance (typically 10^9 Ohms or more), this being a mandatory characteristic for both such measurements. With these two data values, one can identify where the metal is positioned on a Pourbaix diagram, and thus gain insights into the likely behaviour of the metal. However one problem here is that Pourbaix diagrams relate to solution at 25°C. It is possible to re-calculate this data at other temperatures, but it is not a task for the layman.

2.1.12 Oligodynamic Metals in Atmospheric Environments

We have seen that the basis of oligodynamic action is the release of metal ions, either from the metal itself, or a salt of that metal. We have seen how long-established concepts in physical chemistry allow us to predict the rate at which such metal ions will be released into solution. In the case of water-treatment, we are indeed considering a wholly aqueous environment, and concepts such as solubility product apply more or less directly, in these circumstances. But what of the growing application of oligodynamic metals not immersed in water, but used in an everyday environment, as door handles or on work surfaces, where no liquid is present?

To this seeming paradox, there is a simple answer. Water is present. Though it has only really been recognised in recent decades, we now realise that many materials, especially metals, are covered with a thin film of moisture. Invisible to the naked eye as they are, such moisture films are sufficient to allow virtually all the reactions that take place in liquid water to proceed in "air". Just as "Corrosion" is an accepted branch of science and technology, so there is a sub-set of this, known as "Atmospheric corrosion". Another related area of technology is found within the electronics industry. Printed circuit boards consist of metal "tracks" on an insulating plastic base. It has long been recognised in that industry that under many conditions, thin moisture films can form on both the metal tracks and the adjacent plastic, and that various undesirable phenomena can then take place, where ions migrate and react in these thin moisture films. Atmospheric corrosion is too extensive subject in its own right to summarise here. However a key issue is the "Dew Point" - the temperature below which (depending on the relative humidity), a "liquid" moisture film can form. Although in many applications of materials and surfaces carrying oligodynamic metals, the necessary moisture is supplied by perspiration, it is, as far as our understanding takes us, probable that in very hot but dry climates, the oligodynamic action of metals would be significantly reduced.

2.1.13 Oligodynamic Surfaces in Action

It is now widely recognised that germs can be transmitted by manual contact with door handles, taps or other surfaces. Recent studies have shown that sneezing creates aerosols and droplets with a quite astonishing range and velocity, which may well be deposited on whatever surface is to hand. In institutional buildings such as hospitals, doctors' surgeries or nurseries, where infections are more commonly found, and where, by virtue of open wounds or compromised immune systems, people can be more vulnerable, the problem is most serious. Public toilets and food-preparation facilities, especially where poor hygiene standards prevail, are equally at risk. It is known that the persistence of micro-organisms is greatest on organic materials, such as wood, plastics, painted or powder-coated surfaces. As will be seen below, some metals such as aluminium (anodised or not) and stainless steel, behave similarly. As we shall see, it is ironic that stainless steel, which carries in the public mind, connotations of optimal hygiene, shows up poorly in comparative tests, as seen below. From what has gone before, it will be obvious that the presence of oligodynamic metals on any surfaces subject to human contact, will reduce the number of germs on the surface when compared with surfaces where such metals are not present. The thickness at which such metals is present is totally immaterial, from 0.01μm upwards, other than the fact that such very thin coatings would normally begin to wear away within weeks, if not days, of regular contact.

In order to enhance the corrosion rate of silver or copper, as previously noted, a so-called galvanic couple must be set up, where the oligodynamic metal is in close electrical contact with a more noble metal, at which oxygen (from air) is reduced. In the case of silver, this has to be gold, platinum or other precious metal or an alloy of these The more noble metal must, however, be physically discrete. A single-phase silver-gold alloy is not capable of forming a galvanic couple, at least not until, after selective dissolution, where the silver leaches out, pure gold remains in contact with the original alloy. When silver and gold are codeposited, they will normally form such an alloy. However it is possible (123) to deposit the two metals in such a way that relatively pure gold crystals are present in the deposited coating, together with a matrix or relatively pure silver. These are ideal conditions for creating a galvanic couple, where controlled release of silver ions takes place. Similar behaviour is found with a 50% copper-tin alloy, commercially available as "Miralloy" which has been patented (91, 92).

Table 2.1 (a) & (b) Studies on the residence time of micro-organisms on a range of surfaces (from DE 4324994)

	Stainless steel:	Anodised Al:	Powder coated Al:	Brass +10μm Ni:	Brass +10μm Ag;	Brass +10μm Cu/Sn
Cell count (Control)	30,000	30,000	30,000	30,000	30,000	30,000
Staph. Aur. 1hr	1,000	9	1,000	10,000	10,000	1
" " 6hr	–	–	4	5,000	1,000	–
" " 24hr	–	–	1	8	2	–
Staph EPI 1hr	20,000	10,000	1,000	10,000	1,000	–
" " 6hr	10,000	100	100	5,000	100	–
" " 24hr	–	–	10	1,000	3	–
Enterococcus 1hr	1,000	500	1,000	30,000	1,000	100
" " 6hr	1	500	50	10,000	100	100
" " 24hr	–	500	10	1,000	4	–

Table 2.1 (b)

	Stainless steel:	Anodised Al:	Powder coated Al:	Brass + 10μm Ni:	Brass + 10μm Ag;	Brass + 10μmCu/Sn
Cell count (Control)	3,000.00	3,000.000	3,000,000	3,000,000	3,000,000	3,000,000
Staph. Aur. 1hr	10,000	1,000	2,000,000	200,000	10,000	3
" " 6hr	-	1,000	1,000,000	100,000	10,000	–
" " 24hr	-	10	1,000,000	50,000	100	–
Staph EPI 1hr	30,000	1,000	2,000,000	100,000	200,00	1
" " 6hr	1,000	1,000	1,000,000	100,000	100,00	–
" " 24hr	-	100	1,000,000	100,000	100	–
Enterococcus 1hr	10,000	2,000	2,000,000	200,000	10,000	1,000
" " 6hr	1,000	1,000	1,000,000	200,000	10,000	–
" " 24hr	100	50	2,000,000	100,000	1,000	–

Tables 2.1 a & b (9) present data for brass surfaces which have been electroplated with silver, or copper-tin alloy (Miralloy). Results on several other surfaces are included for comparison. The initial colony count was 30,000 or 3 million. Colony numbers are shown after 1hr, 6hrs and 24hrs. Test species were *staph aur., staph epi.* and *enterococcus*. The results in the case of silver, though not unimpressive, are inferior to those from the copper-tin alloy and this is attributed to a lower rate of ion release in the former case, from the pure metal. Had the silver been codeposited, as described above, with another metal in a form allowing a galvanic couple to be formed, the outcome would certainly have been even more pronounced.

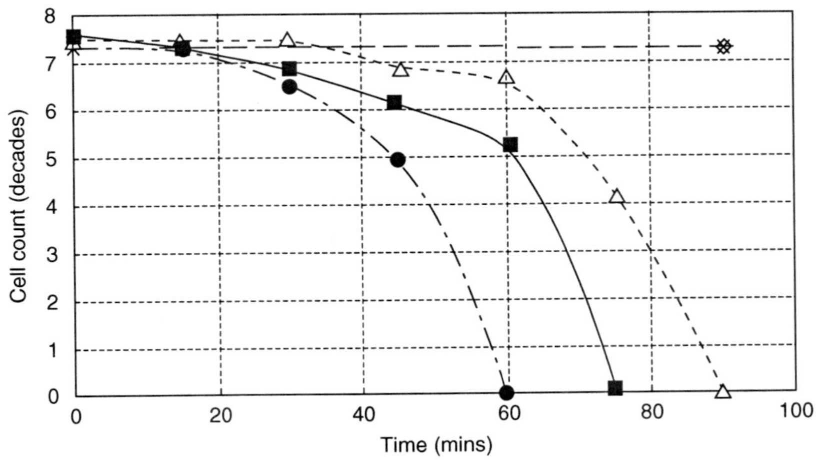

Figure 2.3 Survival of *Listeria monocytogenes* Scott A on various metal surfaces. From top to bottom, stainless steel, silver-nickel alloy, copper-nickel alloy, pure copper (adapted from Keevil (173)).

Numerous studies similar to the above have been reported, notably by Keevil and co-workers (173). Figure 2.3 shows (in idealised form) typical data obtained by him. In this graph, stainless steel appears to show no bactericidal activity at all. In fact, as Keevil shows in his paper, there appears to be an induction period and after 3 hours, the cell count does begin to decline.

In other studies, hygiene researchers in Bochum (Germany) found that in some 50% of institutional building monitored by them, infectious micro-organisms were present at high concentrations, notably *legionella* (124). These and similar species incubate especially in air-conditioning plant and hot water systems. Where warm water remains static for extended periods of time (no water drawn-off), matters are further aggravated. The *legionella* micro-organism, which is fatal in up to 50% of cases, is named after the 30 American Legion veterans who died in 1976, after a re-union in a Philadelphia hotel, in the first recorded instance of the outbreak. In that case, the vector is believed to have been infected moisture droplets in the chilled air circulating through the hotel rooms. Since then, there have been many further fatalities. However in most of these, the infection route is a different one. Rooftop evaporative cooling towers are part of the secondary coolant circulation system, and it is this water which, if not properly sterilised, provides ideal incubation conditions for the bacteria. This water is aerosolised, and can drift downwards from the roof of the building, so infecting passers-by in the streets below.

Studies using silver-plated metal meshes, simply contacting the water reservoirs in such installations, were found to be effective in reducing the level of micro-organisms to more acceptable values (125). There is an inherent problem with all air-conditioning plants in that the very cooling of air tends to reduce the air humidity (water removed by condensation) and this is uncomfortable for occupants of the space in question, leading to dry throats and other minor respiratory problems. This can be corrected by deliberate humidification. However it is precisely in such more humid air that infectious micro-organisms flourish and that the lifetime of an aerosol droplet (which may be infected) is prolonged, since it evaporates more slowly.

In the most recent German VDI legislation, the cleaning and disinfection of such installations at regular intervals is mandatory. This includes bactericidal and fungicidal measures. In all such installations, there are zones where condensation is most marked, and it is here that micro-organisms flourish. These include humidifiers, heat-exchangers (chiller side), sound damping areas, chiller pipework, lamellar precipitators. By coating such surfaces with oligodynamic metals, whether a heterogeneous silver or a copper-tin alloys, the risk of micro-organism growth can be minimised (126).

Turning to a rather different topic, the medical profession repeatedly emphasises the importance of ensuring that eyedrop and nosedrop solutions maintain a sterile state. According to 1996 German guidelines, this must be done by use of incorporating some form of preservative. The problem is recognised to be more acute in the case of repeat-dose applicators, since, in use, the solution can be contaminated by suck-back action, with air or liquid or both thus finding its way into the contents.

At present, it appears almost impossible to find a suitable preservative that meets all the necessary criteria, such as not triggering adverse reactions, offering long shelf-life. In some cases, prolonged use is known to have triggered allergies and irritation to mucous membranes. The most commonly used preservative in eyedrops is benzalconium chloride, though other organic preservatives are also used in practice. Such additives can only be tolerated by the user in relatively low concentration which implies that their disinfective power can readily be exhausted, either because of their reaction with foreign micro-organisms, or their reaction with other compounds in the formulation. For this reason, such medicaments are usually time-limited to ca. 4 weeks after initial use.

An entirely different approach is to use multi-dose applicators incorporating silver-plated components, such as steel valves or springs, where

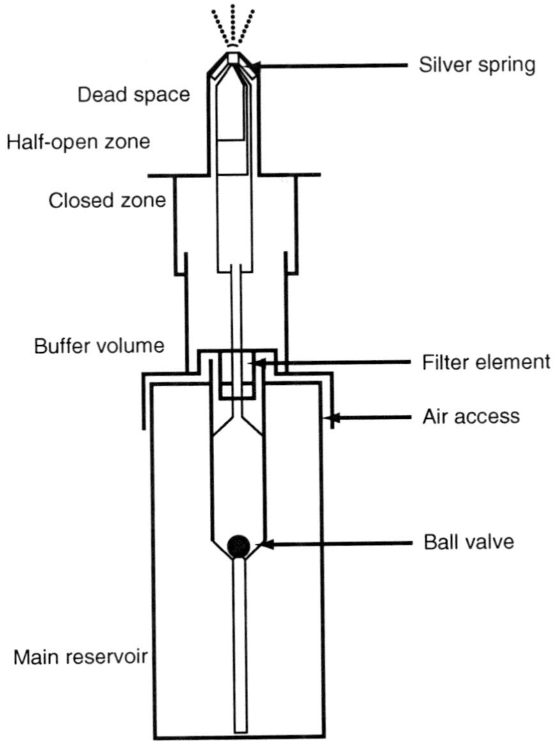

Figure 2.4 Schematic diagram of a specially-constructed multi-shot nebuliser with silver-plated spring and ball-valve.

the oligodynamic action of the silver represses bacterial growth, as shown in Figure 2.4. (24, 25). Trials have shown (26) that bacteria counts such as < 100/ml enterobacteria (ATC 8739); < 10 CBU (Colony Building Units) *staphylococcus aureus* (ATCC 6538P) and *pseudomonas aeruginosa* (ATCC 9027) become undetectable when silver-plated components are used in the applicator.

A surgical problem which has been addressed thanks to the use of silver, relates to procedures after severe skeletal fractures. In some cases, it is necessary to construct an external scaffold, to hold in place the fractured bones during the healing process. As part of this procedure, wires or rods connect the bones to the external scaffolding, necessarily passing through the skin. These locations provide a path for opportunistic infections (42) and antibiotics have proved difficult to use in such cases. (43). Use of silver-coated surgical steel rods appears to offer a useful alternative for prevention of infections, and such rods, PVD-coated with silver, resulted in a decrease in bacterial colony counts or the absence of rod-tract infection, in in-vitro and in-vivo studies (43–45).

In order to assess the biocompatibility of such silver-coated rods as used with external scaffolding, the interaction of silver with tissue cells was studied (46). Neither any genotoxic nor cytotoxic action of silver was detected, and silver-coated implants showed the expected antibacterial action. After four days of cell culture growth, there was excellent cell propagation on the silver coated implants with a higher cell count than on uncoated samples. These results of tissue compatibility with silver surfaces are further supported by studies of tissue reactions on a range of elements and biomaterials (127).

Figure 2.5 shows the elements and alloy mosts commonly found in biomaterials in terms of their corrosion resistance (polarisation resistance) and the type of reaction they induce in a biological environment. In general, a species which is corrosion-resistant in a tissue environment is relatively inert, as for example titanium and its alloys. Toxic effects are seen in elements such as copper, cobalt, nickel and vanadium (apart from allergenic reactions evoked by nickel and cobalt). Between these two extremes, are elements such as molybdenum, iron, aluminium, gold and silver which tend to provoke formation of a hard tissue capsule around them.

Although surgeons are fully aware of problems of post-operative infection and adopt various measures to combat this, such as inclusion of antibiotics at the wound site, such infections are typically running

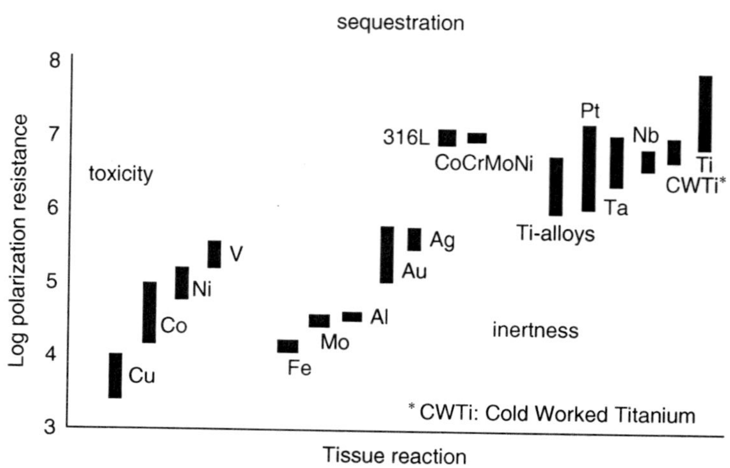

Figure 2.5 Tissue reactions to various elements and biomaterials (127).

at 5% to 35% of all cases. Recognising the antibacterial potential of silver-coated medical appliances, a study of the efficacy of these was carried out, at the same time monitoring any side-effects (172).

In the first trial, 30 rabbits were infected with *staphylococcus aureus*, while 15 animals were fitted with titanium prostheses and another 15 with silver-coated titanium prostheses. In a further study, a search was made for possible toxic side-effects in 10 animals fitted with silver-coated prostheses. In the case of the straightforward titanium prostheses, a 47% infection rate was recorded whereas with the silver-coated titanium implants, this fell to 7%. Various measurements showed a significantly reduced inflammation in the latter group. Analysis for silver ion concentration in blood and vital organs revealed an increased concentration of the metal but with no pathological changes in either the laboratory parameters or histological changes of the organs. From this study it was concluded that silver-coated prostheses do indeed lead to greatly reduced infection rates but with no adverse toxicological side effects and that such implants with their antimicrobial activity, are a powerful tool in reducing post-operative infections.

In contrast to the findings of Bosetti et al (46) who studied the behaviour of silver-coated surfaces, the literature consensus suggests that silver does have a certain cytotoxicity (128). Further studies indicate that after implantation, silver undergoes a chemical reaction and can form highly thrombogenic silver nitrate. $AgNO_3$ has been used to initiate thrombosis in model studies (129). Because of its probable thrombotic action, a direct contact of silver with blood is counter-indicated.

2.2 SILVER DISPERSIONS IN STAINLESS STEEL

The Japanese, like all other developed countries, have suffered from outbreaks of bacterial infection on a more-than-individual scale. Thus in 1996, there was an episode of *Escherichia coli* (0–157) food poisoning, another due to *Staphylococcus aureus* at a later date, and repeated outbreaks of hospital-acquired multi-resistant *Staphylococcus aureus* in 2000. As a consequence of this, the government working with industry, developed a range of products with bactericidal properties, some of which are described below.

As noted previously, (see also below), though stainless steel has a positive public image associated with high standards of hygiene and cleanliness, it is in fact one of the least active in terms of its oligodynamic action. In part, the very corrosion resistance which makes it so attractive, implies that the rate of metal ion release from its surface is extremely low. (27). There have been various attempts to provide a bactericidal functionality, by application of bio-active coatings. All of these are liable to be worn off, either by mechanical action, various forms of wear or as a result of repeated washing (28). A different approach was based on development of alloys which, in addition to iron, nickel, chromium and molybdenum (with other minor constituents) also included an oligodynamic metal such as copper. One such alloy contained 1.5 to 4 wt% Cu, and did indeed display oligodynamic activity. This came, however, especially at the higher copper concentrations, at the expense of corrosion resistance.

Subsequent attempts, based on the use of silver, were more successful. The usual bacteria strains were employed, such *as Escherichia coli, E-coli 0–157, Staphylococcus aureus*, including the multi-resistant strain, as well as *Salmonella*. (29).

Given that the solubility of silver in α-iron is extremely small, of order 0.0002At% (30), alloys were developed in which extremely fine particles of metallic silver are uniformly dispersed in the stainless steel matrix, at a concentration of around 0.04wt% (29). Such steels were formed both as austenitic and ferritic alloys (Table 2.2).

These alloys were formed by continuous casting, then cold-rolled to form 0.7mm thick sheet. Further treatments included hot-rolling, annealing, cold-rolling, recrystallising tempering and acid etching. They were then subjected to a range of corrosion and bactericidal tests. Figure 2.6 shows the results of some of the latter tests. For all bacterial strains used a > 99% cell count reduction in 24 hours was observed, using the appropriate Japanese Standard test (31). The CD-ROM database has numerous abstracts relating to this type of steel.

Table 2.2 Composition of some Silver-Containing Stainless Steels

Composition (wt%)	C	Si	Mn	P	Cr	Ni	Ag
Ag-containing austenitic steel	0,05	0,30	1,00	0,03	18,2	8,5	0,042
Ag-containing ferritic steel (1)	0,06	0,25	0,65	0,03	16,2	–	0,039
Ag-containing ferritic steel (2)	0,01	0,25	0,45	0,03	17,7	–	0,040
Conventional stainless steel, 304 (for comparison)	0,06	0,31	1,02	0,03	18,1	8,5	

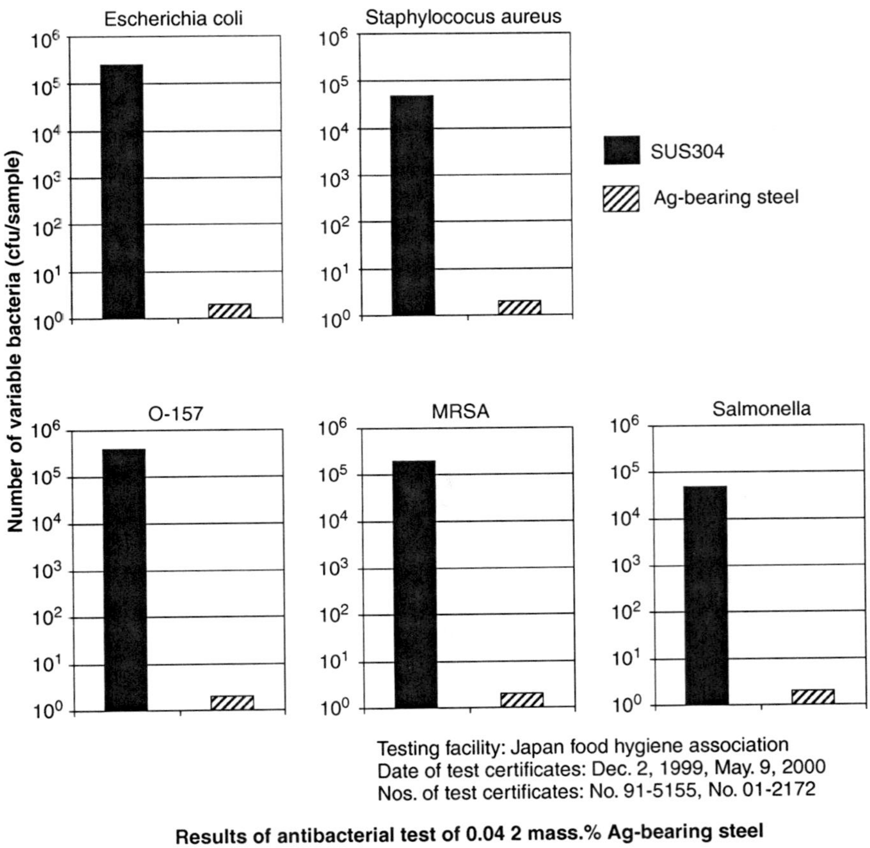

Figure 2.6 Results of bactericidal activity tests for silver-containing stainless steels (0.04 wt% Ag) (29).

A reduction in cell count was observed after 1 hr, with a >99% reduction of after 3 hours. As a control, normal SUS 304 was used. In this case, cell count increased with time, reaching 10^6 CFU after 24 hrs. In corrosion tests, the incorporation of 0.04wt% silver appeared not to influence the corrosion behaviour of the stainless steel alloys. The tests included anodic polarisation in 5% sulphuric acid, salt spray corrosion test in 5% NaCl for 30min, 1hr drying at 60°C, and 1hr at >95% R.H moist air. The cycle was repeated 20 times.

Today, these alloys are widely and successfully used in a wide range of sanitary applications, including laundries, kitchens and other food preparation facilities, hospitals and a range of workplace settings (29).

2.3 SILVER ON AND IN PLASTICS

Plastics are equally commonplace in the food preparation and handling industries, in the home and in industry. In many of these cases, the plastic is subject to frequent handling or contact with the skin. A few examples include door furniture and handles, tools and utensils and more intimate contact such as with ear-muffs and noise protection equipment, hearing aids or plastic safety gloves. Few studies have been reported to determine the extent of infections transmitted by contact with such items, but to discount this type of pathway out of hand, would be premature.

In medicine, as in most other areas of day-to-day life, plastics are increasingly used, both for appliances as well as for working surfaces. Clinicians are all too aware that this has been accompanied by an increase in so-called "polymer-associated infections". Patients, not only those with open wounds, but also those with compromised immune systems, are at far greater risk of infection than the populace at large. Many medical procedures (both diagnostic and therapeutic) are invasive, including catheters, drain tubes, cardio-electric devices etc. In all such cases, there is a wound-site where infection is facilitated, whether by direct ingress of pathogens or their transport on the surface of medical devices made of plastic. Thus, it is known that bacteria can colonise a plastic surface and, by forming a biofilm there, acquire some protection from antibiotics (32–35).

In the treatment of critically-ill patients, the use of central veinous catheters is now an established technique. They are routinely used for infusions, such as saline drips, blood products, and drugs including those used in chronic pain relief. Despite their value in these and many other treatments, such catheters are known to be potential carriers of bacterial or fungal infections. The worst-case scenario is arguably infection of the bloodstream (septicaemia). The dangers and prevalence of hospital-acquired bacterial infection feature daily in the news. Many of these can form, within 12 hours or so, a thick biofilm which defends them from the action of antibiotics. Up to 15% of hospital patients can be afflicted, in varying degree, by such infections. Recent estimates for Germany run at around 75,000 p.a, of which up to 4000 prove fatal (36). Quite apart from the added suffering of the patients in question, such infections almost invariably extend the hospitalisation time, with the cost-implications this brings. Further costs arise when a complete hospital ward has to be closed for the most drastic of disinfection procedures. According to one study by the University of Freiburg, catheter infections equate to some 12,000 Euro per infected patient (36).

Figure 2.7 illustrates a typical infusion system feeding a major vein. The exact configuration of such systems depends on the procedure, but typically features one or more catheter entry-points, 3-way valves, stopcock groups, tubing plugs and connectors, sterile saline solution bags, means for injecting additional drugs into these and air-bleed points. Many of these are possible points of entry for infectious micro-organisms (137)

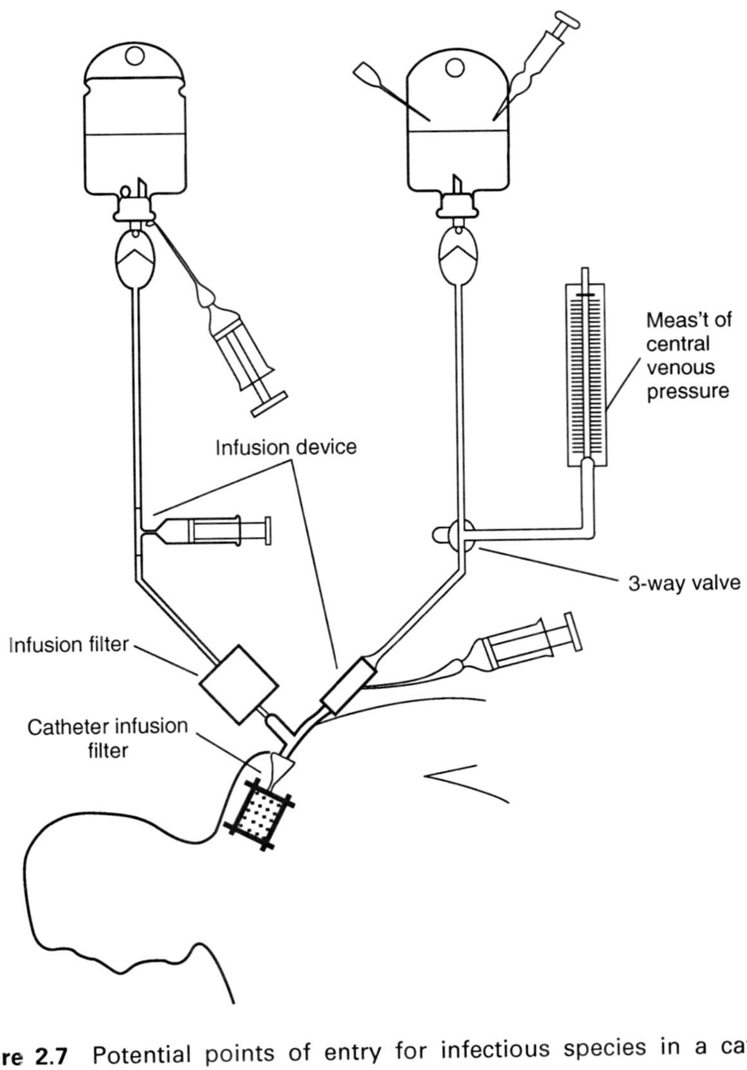

Figure 2.7 Potential points of entry for infectious species in a catheter feed system.

There are other potential danger-points in this or similar types of medical equipment, where use of silver has proved effective in arresting pathogen growth. Fig. 2.8 shows a coupling between a drip infusion system and a vein-dwelling catheter. The radial space between the conical ("male") and the screw-threaded ("female") parts of the coupling provides a small reservoir where drip solution or blood residues can accumulate, to form an ideal breeding medium for micro-organisms (91)

There are various means by which a silver surface can be applied to the transdermal region of a catheter. Thus, for example, a specially-constructed silver annulus can be incorporated around the tubular section of the catheter at the skin-entry point (Fig. 2.9), to provide a flux of silver ions at the point of entry (37, 38)

Catheters can be silver coated (usually only over certain areas) using any of the coating methods previously described in the text or in patents abstracted on the CD-ROM. A more sophisticated approach is the use of the so-called silver iontophoretic catheter (SIC) (39) as shown in Fig 2.10. Two silver wires or tapes (diameter ca. 0.25mm) are spirally wound, parallel to one another, but not in direct contact, around the actual catheter These are then connected to a voltage source, typically 1.5 to 3V, so that they act as electrodes. One wire would then act as cathode, the other as an anode. At the latter, silver would anodically

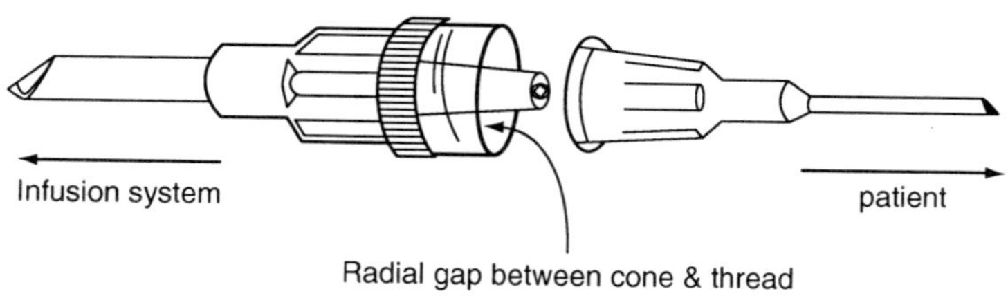

Figure 2.8 Sketch showing coupling between infusion system and vein-dwelling catheter (90).

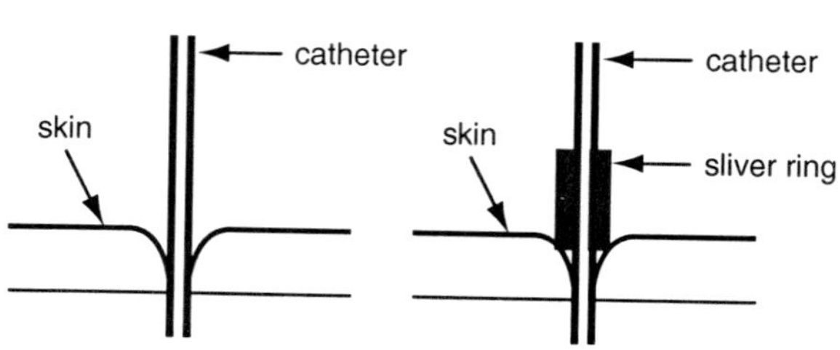

Figure 2.9 Catheter with silver annulus (37, 38).

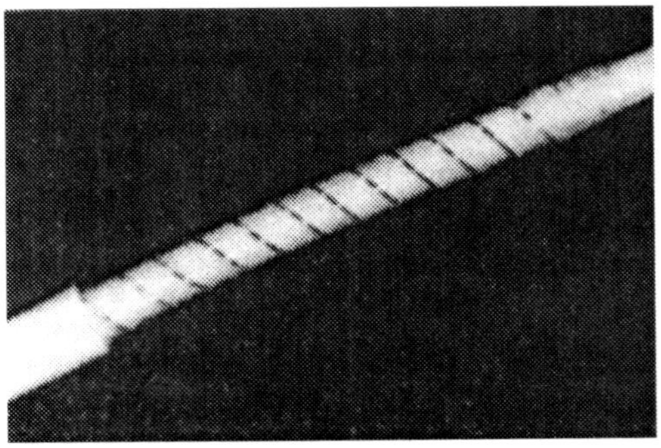

Figure 2.10 Section of a silver iontophoretic catheter.

dissolve, releasing silver ions into the surrounding tissue. An alternative arrangement would be to use one silver wire or tape, and make the catheter body the cathode. The voltage of the power source is adjusted so that a current of 20μA flows and a typical figure for silver ion release rate would be 1600μg/hr. Using this type of catheter, its efficacy was tested against *Staphylococcus epidermis* (the most commonly encountered bacterial infection in catheter use), *Staphylococcus aureus* (a much more virulent strain leading to severe and complicated catheter infections) and *Candida albicans* (the third most common source of catheter infections) and *Acinetobacter baumanii* (a resistant Gram-negative organism found in hospitals). The bactericidal action of this design of catheter was still evident after 4 months.

Another type of catheter featured a PVD silver coating, but trials with this yielded equivocal results (40). In order to promote release of silver ions, a catheter was tested in which both silver and platinum were deposited by PVD onto a plastic substrate (polyurethane or silicone rubber), thereby creating a galvanic cell (see page 66) (10) where silver, being more active than platinum, was released into solution. The presence of 1% Pt in such a coated device made these catheters significantly more active to *Staphylococcus epidermis* than those with pure silver coatings, but without affecting cytotoxicity. That such platinum additions increased the rate of silver ion release, was confirmed by electrochemical methods (chronoamperometry).

A rather different approach was used at Erlangen University where nano-sized particles of metallic silver were used to impregnate the polyurethane catheter. The so-called Erlanger silver catheter is described in (41). Billions of these particles are uniformly distributed in the polymeric matrix. The manufacturing process begins by applying a PVD

Table 2.3 Properties of some polymers used

Code	Description	Density ($gm.cm^3$)	Crystallinity (%)	Melting Pt (°C)
PE-LD	Polythene LD	0.917–0.921	40–50	108
PE-HD	Polythene HD	0.950–0.954	60–80	131
PP	Polypropylene	0.905–0.909	60–70	160–165

coating of silver to the thermoplastic polyurethane. Granules of this material are then extruded in a double-helix screw extruder, in order to homogenise them, and are then pelletised. The 20–80nm silver particles are uniformly distributed in the polymeric matrix, at a concentration not exceeding 2wt%. As a result of this sequence, a composite is formed where there are no metallic silver particles at the actual surface, since these could induce adverse blood coagulation cascading (41, 129).

Aromatic and aliphatic substituted polyurethanes are hygroscopic and able to take up both water and electrolyte. Water uptake is higher in the case of barium sulphate-filled polyurethane, and the polymer can end up, in effect, as a sponge, allowing water and electrolyte access to three-dimensionally-distributed silver. Calculations show that at 0.8wt% silver/gm of polyurethane, the Erlanger catheter contains 1.5×10^{15} silver particles of 50nm diameter with an equivalent surface area of $0.45m^2$/gm of the polymer. Electrochemical experiments (reverse voltammetry) indicated a silver elution rate of 1μg Ag/24 hrs per meter catheter length (internal catheter diameter = 1.8mm), thereby allowing a steady release of silver ions for many months.

A later development in infusion technology was based on the use of silver chloride distributed in a polyurethane or polypropylene matrix, formed by extrusion (90). The materials used are summarised in Table 2.3. Various mixtures were tested with 1 to 2.5 wt% AgCl.

All of these exhibited a release of silver ions, and polymers with higher degree of crystallinity were found to release silver ions more rapidly than those of lower crystallinity. Microbiological tests gave somewhat conflicting results and these were attributed to liberation of chlorine gas as a result of photochemical action of daylight on the silver chloride (90). It was found that incorporation of TiO_2 mitigated this (TiO_2 at 10% concentration) and furnace black (PrintexL) at 2% achieved a similar effect.

3

Nano-Silver Particles in Gels, Paints & Textiles

3.1 GENERAL COMMENTS

Increasingly, paints and other organic coatings are seen as more than simply protective or decorative covering materials – they also offer some other functionality. Easy-cleaning, soil-repellent (non-stick) and bactericidal surfaces are ever more prized, and organic coatings are now available which offer some, if not all of these attributes. Incorporation of silver nano-particles, typically 5 to 10nm in size, in a polymeric matrix, leads to coatings with bactericidal and fungicidal activity. Numerous articles have discussed such coatings which provide a controlled release of silver ions.

As with previously-discussed practical applications of oligodynamic action, there is a choice between use of metallic silver or one of its sparingly-soluble salts. In one instance (50), composite particles were formed of around 20 to 30wt% AgCl on the external surface of a titania support. The AgCl is specially formed in the surface of the highly porous titania particles, of size 0.3 to 5μm, as shown in Fig. 3.1. In aqueous solution, these would typically form a concentration of 20–40ppm silver ions. The following equations describe the situation:

$$AgCl_{solid} <===> AgCl_{dissol'd} <===> Ag^{+} + Cl^{-}$$

These three stages are in equilibrium, which is to say that, in a closed system, once a certain concentration of Ag^{+} and Cl^{-} ions is reached, (as determined by the solubility product), no more dissolved AgCl will dissociate, and no more solid AgCl will dissolve into solution. In practice, it is most unlikely that equilibrium will be reached. The Ag^{+} ions will be taken up by micro-organisms present, and any flow of water will remove all dissolved species, causing more solid AgCl to dissolve. In seawater, a lower concentration of silver ions is found, typically 2 to 4 ppm. This is a consequence of the solubility laws. As we saw earlier, the solubility product, L_{AgCl} can be expressed as:

$$L_{AgCl} = [Ag^+]\,[Cl^-]$$

Because of the high concentrations of Cl^- ions in sea water, if the condition of this equation is to be met, as $[Cl^-]$, the chloride ion increases,

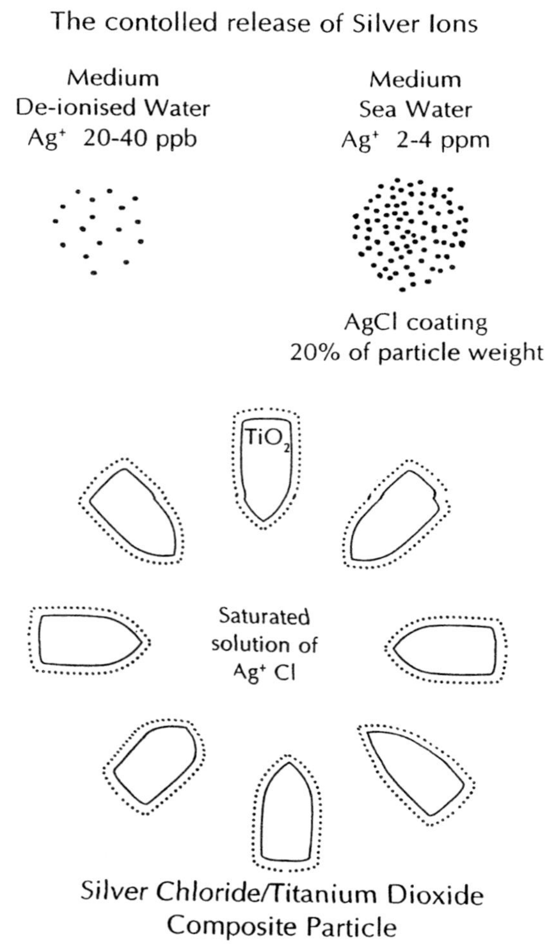

Figure 3.1 Silver chloride – titania composite particles.

so $[Ag^+]$, the silver ion concentration, must decrease, since L_{AgCl} is a constant.

By suspending these composite particles in an aqueous medium or a surface-active gel, the active silver-ion releasing particles can readily be brought into contact with whatever pathogen-containing species might be present. A widely used vehicle in pharmaceuticals and cosmetics and medical preparations, is sodium dioctyl sulphosuccinate, also used as a latex emulsion and in toiletry preparations. Use of such vehicles improves the controlled-release of the silver ions. Such formulations are entirely harmless to humans. Of the three constituents, silver chloride is sparingly soluble in aqueous media and passes through the intestinal system with minimal uptake. TiO_2, titania, is widely used in cosmetics and colorants and is believed to be completely inert, while the sulphosuccinate is likewise widely used in pharmaceutical and cosmetic formulations.

DuPont Powder Coatings produce a powder coating, trade-name Alesta® in which silver, in ionised form, is bound to a biocompatible zeolite. Alesta® is applied in the same way as conventional powder coatings and, being mainly inorganic, are thermally stable up to around 500°C and can be used in the pH range 3 to 10. (1b)

Silver compounds, able to ionise, display a biocidal action even when incorporated in a polymer matrix, provided there is a reasonable degree of porosity. The same is true of silver in colloidal form, incorporated in a matrix. In both cases, a sufficiently high diffusion rate through the matrix to the surface, is necessary. The downside of this approach is that given such relatively high diffusion rates, the silver, whether in ionic form or as a colloid, is soon exhausted, especially when the surface is permanently in contact with solution or subjected to repeated washing.

Colloidal silver can be incorporated into coating formulations by addition and mixing (56–58). However there is always the danger that the colloid will break down and coalesce (coagulate). The stability of such colloids is contingent on each constituent particle retaining a charge of the same polarity as that on all the other particles, so that coalescence is prevented by mutual charge repulsion. If these charges are in any way neutralised, coagulation ensues. In the main, protonic solvents are required to avoid charge neutralisation, and the colloids will often retain their state only within a limited pH range. These conditions restricts the number and type of coating formulations that can be used in this approach.

Using the sol-gel approach (51–53), colloidal silver can be formed in-situ at low to ambient temperatures. To ensure controlled-release over longer periods, larger colloidal silver particles are called for. These can be incorporated in a paint coating which has a relatively high elasticity. Equally they can be used on flexible substrates, such as textiles.

Typical coating formulations would include a) a hydrolysate or condensate of at least one hydrolysable silane ($R_aSiX_{(4-a)}$) and at least one non-hydrolysable (carbon-containing) substituent and b) a silver compound (preferably as a complex (51)). To ensure reasonable scratch-resistance and hardness, inorganic nanoparticles such as silica, alumina or ITO (indium-tin-oxide) or zirconia can be added. Various solvent additions might also be used.

The hydrolysis or condensation process mentioned above is carried out either in absence of a solvent or preferably in an aqueous or aqueous-organic reaction medium, optionally in presence of an acidic or alkaline condensation catalyst such as HCl, HNO_3 or ammonia. This results in a partial hydrolysis or (poly)condensation of hydrolysable species (condensation precursors). The degree of condensation, and thus the viscosity, can be varied by selection of a particular solvent. The resulting sol is used to formulate the coating formulation, with the silver-containing compound(s) and other components then being added. The coating is then applied in any of the usual ways (painting, spraying, pouring, roll-coating, centrifugal coating, meniscus coating, rake-coating or electrocoating). Conditions in each case are adjusted to give the desired dry-film thickness (1 to 15μm). This is followed by drying at $< 40°C$ and stoving (50 to 100°C) or UV-curing, in order to form the colloidal silver. Larger-sized colloidal silver particles (10 to 30nm) are formed specially rapidly under UV radiation. In the absence of UV irradiation, smaller silver colloid particles form during stoving (5 to 20nm). These procedures offer an elegant means of avoiding a need to form the silver particles in a separate stage, with the associated compounding this might require.

Typical hardening times at temperatures $< 130°C$ are 2 hours or more. In the TEM image shown in Fig. 3.2 (12), the nanoparticles of silver can be identified. Fig. 3.3 shows how, after an induction period of some 15 days, a constant rate of silver ion release is observed. Thus, using a silver-loaded paint with 0.87wt% Ag, as compared with an uncoated polycarbonate surface, formation of a biofilm based on *Escherichia coli* bacteria was largely suppressed

Silver-loaded paints of the type described above, can be applied to virtually any material – metal, glass, ceramic, plastic, paper, wood. Given the modest stoving temperatures required, use of heat-sensitive

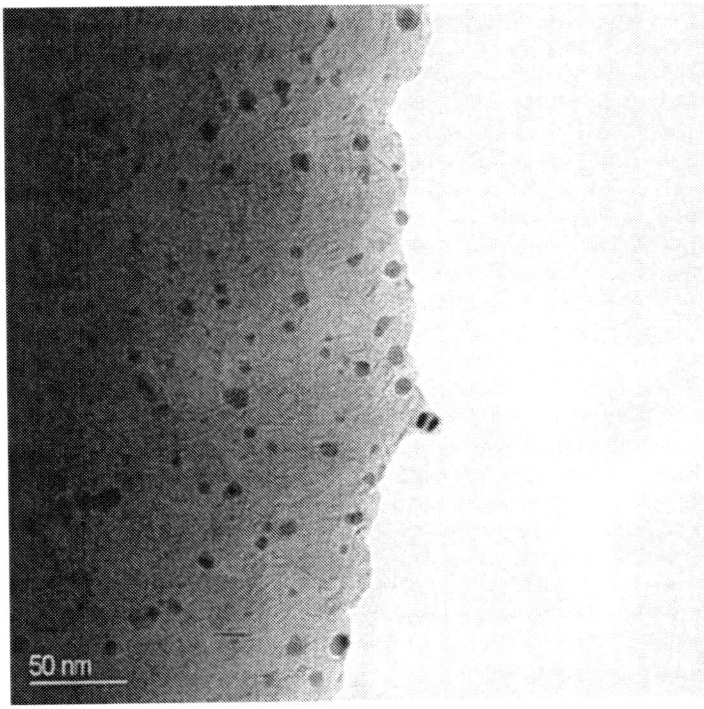

Figure 3.2 Transmision Electron Micrograph (TEM) image showing nanoparticles of silver in a UV-cured organic-inorganic composite coating.

substrates need not be a problem. It might be noted that this type of coating is substantially transparent.

The silver-loaded paints described above have a wide range of applications. They can be used in powder-coating of metallic surfaces, on switchgear cabinets or work surfaces, where their scratch-resistance and bactericidal properties can be important. Equipment and appliances for food preparation, transport and storage are another important application area (refrigerators, freezers, containers, carbonated water or drinks containers and industrial equipment such as food conveyor belts.) Fig. 3.5 shows an example of this. Large-scale catering facilities and meat-handling and food-packing plants are other areas of use. Also a wide range of door furniture, cabinet handles, WC fittings, taps, bathroom appliances such as toothbrush handles etc. Fig. 3.6 shows earplugs coated with silver-loaded paint. Architectural and related fittings in airports, railroad termini, clinics and hospitals, works canteens are other examples. Similarly-formulated solutions can be used to produce

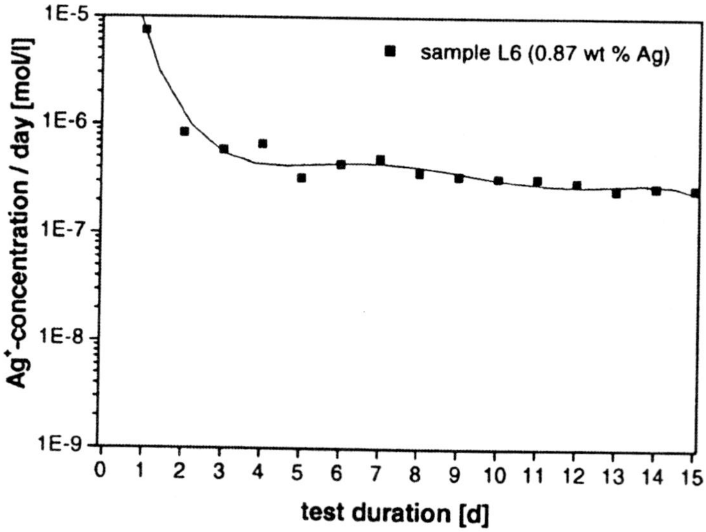

Figure 3.3 Time-dependant release of silver ions from a UV-cured composite coating (as in Fig. 3.2) into a physiological solution.

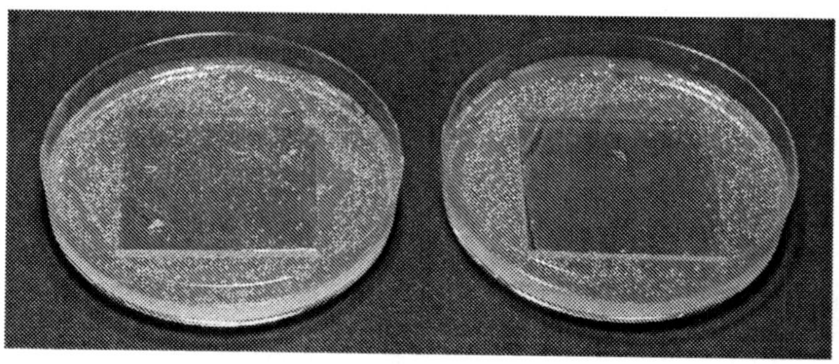

Figure 3.4 Cell growth of *Escherichia coli* bacteria on an untreated surface (left) and one with a silver-containing coating.

silver-loaded textiles and fabrics, where solid silver-containing particles adhere to individual fibres.

The medical profession is well aware of the main vectors for transmission of bacterial and fungicidal infection, be these through contact with the skin or as airborne aerosols. Wherever large numbers of people meet or pass one another and/or where meat and fish are handled and in hospitals or doctors' surgeries – there are locations where the risk of infection is at its greatest and where silver or copper-coated surfaces have an important role to play in stemming the transmission of infection.

Figure 3.5 Silver-coated roller-conveyors in a food-preparation plant.

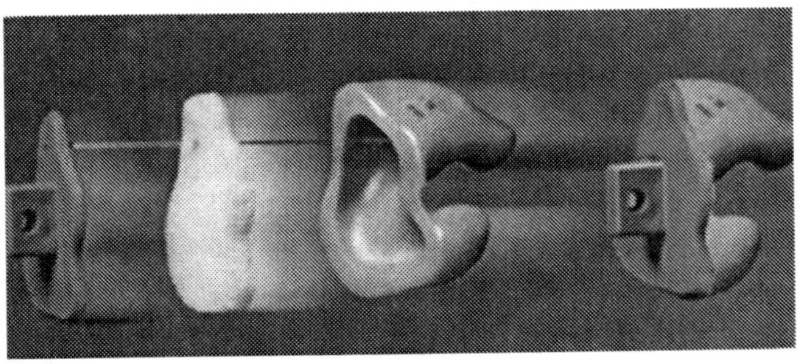

Figure 3.6 Ear-plugs and hearing aid inserts coated with silver-loaded paint.

3.2 SILVER-LOADED RESINS FOR DENTAL APPLICATIONS

Dental caries continues to be one of the main causes of poor dental health and is a bacterial disease. While by no means the exclusive cause, the species *Streptococcus mutans, Streptococcus sobrinus* and lacto-bacilla are undoubtedly major culprits in carious attack. As noted in a previous chapter, organic surfaces are specially prone to bacterial growth and biofilm formation. This is equally the case for the resins used in restorative dentistry. Bacterial growth tends to be more vigorous on these surfaces and – though not overly harmful *per se* – can easily spread to adjacent hard or soft tissue and thus induce secondary caries.

Use of silver-loaded glass ionomer dental cements was shown, in vitro to inhibit growth of oral *streptococcus* around fillings made using the material (54). This effect was ascribed to release of silver ions, fluoride ions or both. Initial attempts to inhibit bacterial growth and biofilm formation were based on incorporation of chlorhexidine. However it was soon apparent that this species remained active for only a relatively short time. Various attempts to prolong the activity of this species, brought with them a deterioration of the physical and mechanical properties of the dental cement. These problems were overcome by incorporation of silver in ionised form in particles based on inorganic oxides, having a structure of either a ceramic or a silica-gel. Such particles were then included in the dental cement. Examples of this approach include use of zirconium phosphate [$Zr_2(HPO_4)_3$] or silica gel, as support materials, sometimes designated SSAM (55). Such antibacterial dental cements are typically prepared at the chairside by mixing a powder and liquid. A typical composition of the liquid might be 49.5 wt% triethylene glycol dimethacrylate with 49.5wt% 1,6 bis(methylacryloxy-2-ethoxycarbonylamine) -2,4,4- trimethyl hexane, with 0.5% dl-camphorquinone as a photoinitiator (for radiation curing) and the same amount of 2-(dimethylamino) ethylmethacrylate as photo-accelerator. The powder component would typically be silanised silica and/or SSAM.

The success or otherwise of such resin-based composite coatings, in terms of their bactericidal action, depends very much on their formulation. The hypothesis has been advanced that such effects are related to formation of O_2 or OH^{*-} radicals formed by photochemical action rather than release of silver ions (55)

3.3 SILVER-CONTAINING CERAMICS AND GLAZES

The idea of bringing together the bactericidal properties of silver and the many advantages of ceramic coatings, is far from novel and numerous studies have been carried out with the aim of optimising large-scale production of such materials. Various patents have been filed, such as (59–62) and some of the processes described in these patents are used on an industrial scale.

Creating ceramic materials capable of releasing silver ions is far from easy, since these, including ceramic glazes, normally require firing at around 1200°C. It has been demonstrated that above 470°C, silver ions such as those present in silver nitrate, can co-exist with metallic silver particles (63). The solubility of metallic silver in glass is low (64).

Figure 3.7 shows a proposed design of a T-piece for use in a medical infusion apparatus, such as a saline drip. It was formed from a silver-ion-releasing ceramic material and then tested (90). The idea was to impregnate the porous ceramic with silver chloride in a liquid suspension. Thereafter, in use, the silver ions would then be released.

Table 3.1 shows characteristic data for some commercially available ceramics, used in the above device. Impregnation (also referred to as infiltration) is optimised using the principles of the Hagen-Poiseuille equation (shown below), i.e maximising pressure differential, extending immersion time and minimising the viscosity of the infiltrating solution

The Hagen-Poiseuille equation is:

$$\frac{v}{t}[m^{3\cdot}.s^{-1}] = \frac{\Delta p \cdot \pi \cdot r^4}{8\eta.L}$$

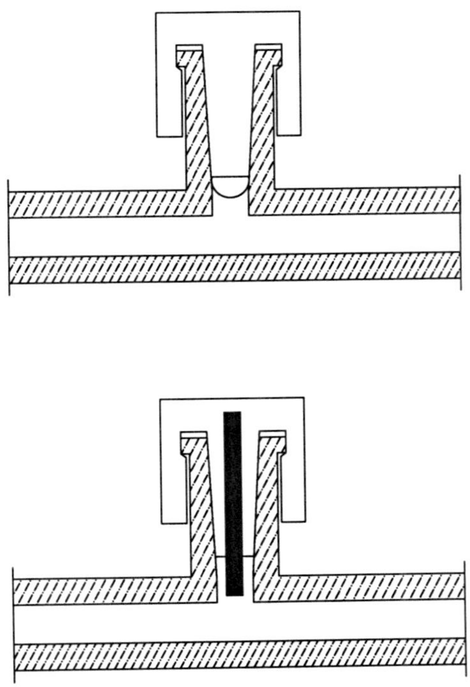

Figure 3.7 T-piece connector with conventional closure (above). With modified closure (below).

Table 3.1 Characteristic parameters of some commercially available ceramics

Designation	Sample Dimensions		Surface Area	
	Diameter (mm)	Length (mm)	Geometric $M^2 \cdot g^{-1}$	True $m^2 \cdot {}^{-1}gm$
Super-Pyrostat C 230	2,20	8,00	$1,1 \cdot 10^{-3}$	1,54
Thomit 124L	2,15	6,80	$8,0 \cdot 10^{-4}$	0,07
Thomit 700	2,50	8,00	$5,7 \cdot 10^{-4}$	0,04
Rubalit 702	2,50	8,00	$3,3 \cdot 10^{-3}$	0,04

where:

V = volume of infiltrating liquid
Δp = pressure difference (head) between level in the capillary tube & the main liquid
r = radius of capillary tube (m)
t = infiltration time (sec)
η = dynamic viscosity of infiltrating liquid (Pa.sec)
L = length of capillary (m)

Porous substrates can also be liquid-impregnated by making use of capillary action, as demonstrated by placing a capillary-bore tube in a beaker of liquid. Thanks to capillary action, the liquid in the capillary bore will rise, typically several centimeters above the level of liquid in the beaker. The relevant expression describing this is:

$$h[m] = \frac{2 \cdot \sigma \cdot \cos\theta}{r \cdot \zeta \cdot g}$$

Where:

σ = surface tension ($N.m^{-1}$)
r = capillary radius (m)
θ = wetting or contact angle
ς = liquid density ($kg.m^{-3}$)
g = gravitational constant ($m.s^{-2}$)
h = capillary rise (height of liquid)

Unfortunately, this approach has to date met with little success. Thus the ceramic Super-Pyrostat C230 lent itself well to infiltration and showed good bactericidal action against *Candida albicans*.However the ceramic was not hard enough for practical use. Harder and more fracture-resistant ceramics such as Rubalit 702 or Thomit 700, proved more difficult to infiltrate and gave no consistent results for this (90).

Following these disappointing results, work was carried out at the Ceramics and Vitreous Enamel Institute at Clausthal Technical University where a specially-developed bioceramic was produced (161) with silver-release and bactericidal properties equal to those shown by Super-Pyrostat C230, but with greatly improved mechanical properties. The ingredients for this are shown in Table 3.2.

The compounded power was isostatically pressed at 20 MPa and the green material sintered at 1500°C (1 hr). Key parameters for the sintered product are given in Table 3.3. (90)

Most recently, glasses have been developed in which silver ion exchange takes place in the glasses, which have a high sodium ion content (Table 3.4)

Glass of the above composition was milled with a mixture of 98% $NaNO_3 + 2\%$ AgO/Ag_2O prior to testing. The following species succumbed to the bactericidal action of the silver-containing

Table 3.2 Materials used to produce Clausthal Bioceramic (Drescher 1995), (161)

Parameter	Value
Chemical composition (wt%)	Al_2O_3 99,8; $Na_2O \leq 0,1$; Fe_2O_3 0,01; SiO_2 0,01; $TiO_2 \leq 0,005$; Ca0,01; $MgO \leq 0,1$
Shrinkage onset (°C)	1100
Water content (wt%)	$1,50 \pm 0,04$
Binder content (wt%)	$0,90 \pm 0,04$
Frit grain primary size range (μm)	0,22–3,00; 50% < 0,72
Agglomerate form	Hollow spheres, partly dented 40–60mm
Specific surface area (sq.m/gm)	11 ± 0.12

Table 3.3 Key Characteristics of the Clausthal Bioceramic after sintering (Drescher 1995) (161)

Comments	Parameter
Shrinkage on firing (% linear)	$15,69 \pm 0,05$
Sinter density ($gm.cm^{-3}$)	$3,66 \pm 0,05$
Vickers Hardness (MPa)	15588 ± 79
% Open Porosity (Vol.%)	$5,37 \pm 0,05$

Table 3.4 Composition of an Ion-Exchanging Glass

Oxide	SiO_2	Al_2O_3	Na_2O	Li_2O
Wt %	72	3	20	5

glass: *Staphylococcus aureus* ATCC 25923, *Escherichia coli* ATCC 29922, *Bacillus cereus* and *Pseudomonas aeruginosa.*

Silver and sodium ions, both monovalent, also have very similar ionic radii and several studies have shown that sodium ions are ideal candidates for substitution by silver (63, 65–67). Ag^+ ions are more strongly polarised than Na^+ and have a greater ionisation potential and are more electronegative (67). This implies that the Ag..O bond has a more pronounced covalent character than its analogue, the Na. . . O bond and it has been suggested that this explains the formation of a strong Ag-O-Si linkage which forms by displacement of the sodium ion.

Following this ion-exchange process, a subsequent thermal treatment may be applied which accentuates the yellowish tinge of the glass. The silver ions diffuse further into the glass matrix and at temperatures > 470°C, silver crystallites begin to form. This conversion of ionic to elemental silver, and its diffusion away from the outer surface of the glass into the bulk, both processes taking place more rapidly as temperature increases from 400 to 600°C, results in a decrease in the bactericidal activity of these glasses (22), confirming once again, that it is the ionic form of the metal that is active. Other workers have used alternative routes to forming of bioactive glasses, such as sol-gel methods to produce a four-component glass (68).

There are several approaches to characterising and defining the bioactivity of a material, but one of these is the ease with which it binds to bone or hard tissue. The binding mechanism involves chemical reactions whereby an outer layer of calcium hydroxyapatite (HA) is formed. The rate of formation of such layers can be directly correlated with the bioactivity of the material involved. Bioglasses are mainly based on SiO_2, CaO and P_2O_5 with additional components such as Na_2O, K_2O, B_2O_3, MgO or fluorides, all of these being known glass-formers.

The rate of HA formation depends not only on the composition of the bioglass, but also the route by which it was formed. Thus glasses formed using sol-gel methods tend to be more active, on account of their high specific surface area and porosity (69) and for this reason, glasses so made are usually selected for partial replacement of sodium ions with those of silver. The four-component glass noted above had the following composition: 76wt% SiO_2, 19wt% CaO, 2wt% P_2O_5 and 3wt% Ag_2O.

The bioactivity of this glass was studied in simulated body fluid (SBF, or physiological solution). Fig. 3.8 a) shows the rate of silver ion release in SBF and as Fig 3.8 b) shows, there is a drastic decrease in the rate of release over the first week. The plot closely resembles those for release of Ca^{++} and phosphate ions. In-vitro bioactivity studies show that the silver-doped glass behaves little differently from its undoped analogue, which is explained in terms of the very small rate of release of silver ions. (68)

The workers here were presumably aware of the fact that the rate of diffusion of silver ions out of the glassy matrix and into solution, obeys Fick's Law of Diffusion. They thus did not repeat the erroneous measurements of Braden et al who, in the 1970's, studied the release of fluoride ions from silicate dental cements into aqueous solutions, apparently not recognising that as the concentration of those ions in solution built up, so the concentration gradient driving the process, decreased.

Silver-loaded microspheres (diameter < 1μm) were also formed using sol-gel procedures. Tetraethoxysilane (TEOS) was first hydrolysed in ethanol, by addition of water, after which aluminium triisopropylene dioxide was added to form Si-O-Al bonds. An ammoniacal silver solution was then added, resulting in formation of silica microspheres (0.4 to 0.6μm diam) which also contain silver and aluminium ions. The nominal compositional range of these is Si/Al/Ag = 1/0.01–0.03/0.003 to 0.03; Al/Ag = 1–3.3. That is to say in terms of stoicheometry, these

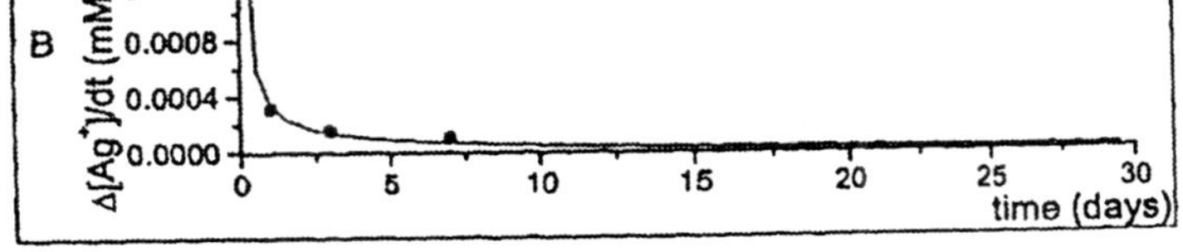

Figure 3.8 A) silver ion release in simulated body fluid. B) rate of silver ion release (i,e first differential of the data in A). The physiological solution was replaced every 24hrs.

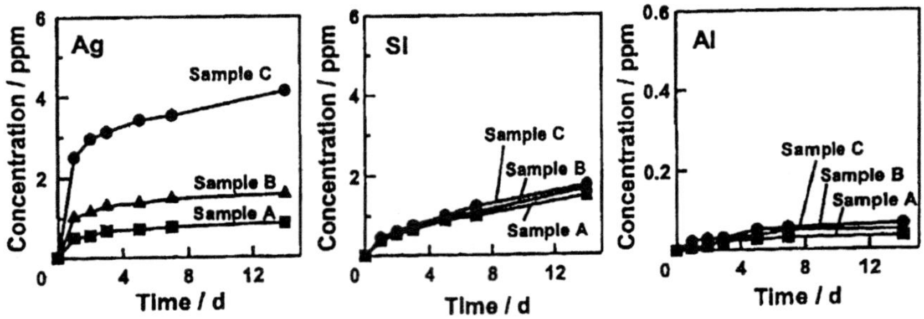

Figure 3.9 Release rates of silver, silicon & aluminium from three samples (A, B, C).

are aluminium (and silver) loaded silicate glasses rather than actual aluminosilicates.

Microspheres of composition Si/Al/Ag = 1.0/0.01/0.01 show excellent bactericidal activity towards *Escherichia coli* (70). Fig. 3.9 shows the release rates for Ag, Si and Al from glasses of three different compositions. The Ag^+ ions are incorporated in the silica lattice in the form of $[AlO_4]^-$ Ag^+, and are liberated into aqueous solution by exchange with hydrated protons H_3O^+.

Figure 3.10 reveals that the rate of silver ion release obeys the classical $\sqrt{t}$ plot, indicative of a diffusion-controlled process, presumably here solid-state diffusion inside the glass.(71)

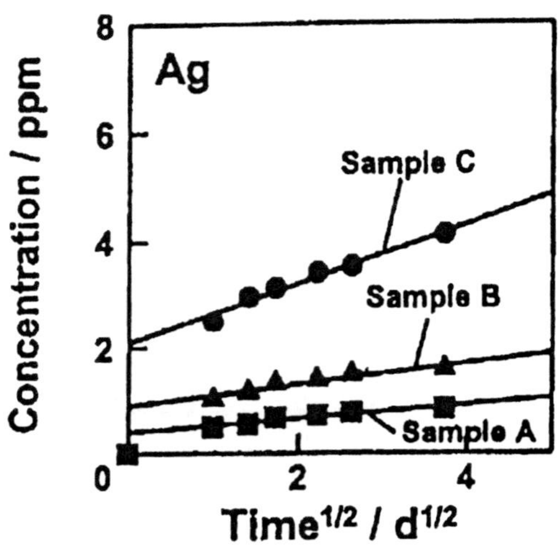

Figure 3.10 Silver ion release plotted vs. $t^{0.5}$.

3.4 SILVER IN TEXTILES OR FABRICS

The time has long passed when textiles served only to make clothing, or domestically-used fabrics for curtains, upholstery and bedding. So-called "technical fabrics" are increasingly used, in many cases being hidden from view. In the automobile industry, seat and roof upholstery is evident enough, but the reinforcing fabrics used in tyres or the battery are rarely exposed to view. The same is true of non-metallic bodywork in cars, boats or airplanes, where resin-impregnated fibre-mats constitute the structure of the composite in question, and also in the building. The same is true in the building trade, where textile moisture barriers are widely employed. In medicine and surgery, the clinician has at his disposal, a huge range of bandages, wound-dressings, plasters. In Germany, it is estimated that some 40% of all woven materials are used in these or other technical applications. Of these, some 96% are based on synthetics rather than natural fibres and development of new polymers, new combinations of polymers and novel composites incorporating woven materials, continues apace.

On the consumer front, there is a demand for clothing and domestic textile materials with properties unknown in the past, such as anti-static cloth, or "effect" materials, whose colour varies as the angle at which it is viewed. The ill-defined phrase "Life-style" captures certain of these aspirations. Women have always been seen as the main purchasers of perfumes and other cosmetics and related products such as anti-perspirants. In recent years, at least in some national markets, this is no longer true. Male body-care products such as after-shave or anti-perspirants are now a market as large, if not larger than that for women. Thanks largely to aggressive advertising campaigns, both men and women are increasingly preoccupied with real (or imaginary) problems of body-odour. New phrases such as "feminine hygiene" have a resonance which need not be spelled out here.

Two developments of the last decade or so, should be identified here. Firstly, a wider recognition (though nothing new to medical or bioscience professionals) that many body odours have their origins in bacterial growth supported by body fluids, of which perspiration is but one. At various creviced locations on the body surface, bacteria find both the warmth and moisture which promotes an explosive growth rate. The skin at these locations is softer than exposed areas, for example, the face or hands, and nutrients thus more easily accessible. Bacteria and fungii can flourish and so generate unpleasant smells from their by-products, similar to those found from infected wounds, rotten fish, meat or decaying vegetable matter. *Staphylococci* release fatty acids and other bacteria release proteins from keratinous tissue

The second development is that of clothing which is capable of actively combating body odour, because the textiles used incorporate a bactericidal action. Socks were perhaps the earliest application of this technology, which now extends to shoes, and underwear and, most recently, casual wear such as football shirts. A somewhat surprising application of this approach has been its use for military clothing. It is recognised that soldiers in combat may lack access to washing facilities over extended periods and the US Army is only one of several such bodies now developing such clothing.

In the light of previous chapters, it comes as no surprise that incorporation of silver, in metallic form or as a compound, into textiles, is a widely-used strategy, though by no means the only one. Some examples of this are found in (72–83, 93–94). See also the CD-ROM database.

Thus in 1990, Wacker Chemie developed a process for bonding cyclodextrin (a ring-type saccharide) onto textile fibres. The structure of the cyclodextrin was modified so that it resembled a certain type of dyestuff. It was found that the cyclodextrin behaved somewhat like a reservoir in that it adsorbed the odour-producing compounds up to the limit of its capacity, then releasing them over time to air. This reversible action appears to have been physical rather than chemical. Another approach was to incorporate organochlorine compounds which served to inhibit bacterial growth.

A detailed study of clothing and textiles featuring bactericidal species was carried out by the Institute for Hospital Hygiene and Environmental Medicine at the University of Freiburg (72). As part of this, they expressed a certain concern at possible adverse side-effects of these fairly potent chemicals in such close proximity to the skin. As will be evident, the use of silver avoids any such concerns.

3.4.1 Metallised Textiles and Fibres

Intuitively, most materials scientists would have severe misgivings about incorporation of silver or silver compounds onto or into textiles or textile fibres. Both from a chemical and a physical viewpoint, the two would seem to be poorly compatible. In chemical terms, without resorting to a silane type bridging, it is hard to envisage a robust bonding mechanism. In physical terms, the lack of stability of a fibre, which both flexes and wears, is unpropitious. Fibres are continuous, silver or silver compound particles are discrete. Despite this unpromising starting point, silver-containing textiles are a commercial reality.

An Italian company, Nylstar has developed a process (Fig.3.11) where fine silver threads are woven in with polyamide yarn. Alternatively they use metallised thread, manufactured by companies such as Kunert or Calida. Trevira is another well-known manufacturer offering bioactive polyester fibres incorporating silver.

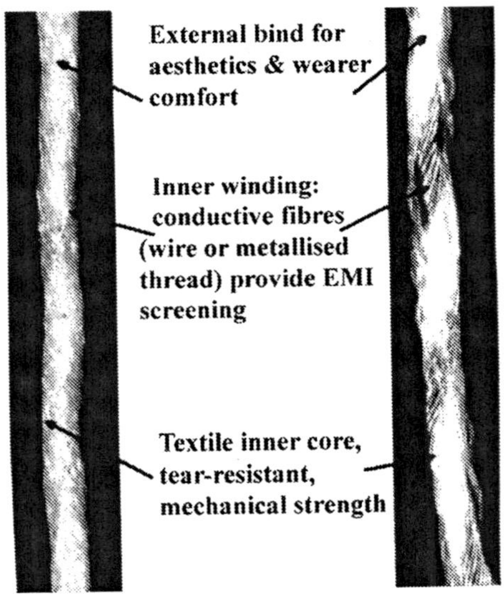

Figure 3.11(a) Silver-coated threads used in woven textiles.

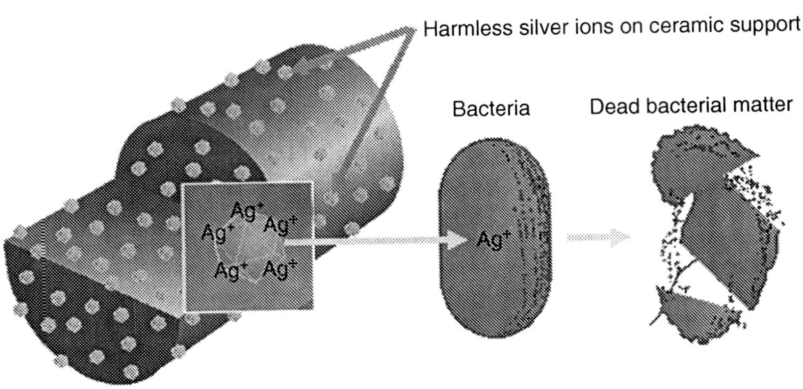

Figure 3.11(b) Silver on a ceramic support (Trevira).

The use of silver-containing fibres in textiles for everyday use, is controversial. There are questions as to whether the results of laboratory tests can be translated to actual use, and also concerns as to the interaction of silver with naturally occurring skin flora. Against this is the undoubted fact that, over thousands of years, there have been no suggestions that silver jewellery worn next to the skin is in any way harmful.

On a more positive note, silver-containing fabrics have been used in treatment of some forms of dermatitis, such as those where the natural flora are affected, as in neurodermatitis (72). Thus, dressings with a very high silver content of order 20% (such as "Padycare") are used to ease irritation caused by neurodermatitis. Such beneficial effects have been confirmed by clinical trials, albeit of only a few days duration. Neurodermatitis can last for many years and there appear to be no studies covering the use of silver dressings for anything like that period of time. [72, 73]

Figure 3.12 shows a schematic cross-section of a micro-fibre and a section of a silver-containing woven fabric (74). The means by which the silver was incorporated is not disclosed.

Manufacture of anti-bacterial polyester fleece (PET-fleece) involves use of nano-silver colloids. These are formed by chemical reduction of silver nitrate (0.5M) with 0.1M hydrazine in the presence of a wetting agent (similar to conditions used to form silver mirrors). The equation can be written:

$$N_2H_4 + 4Ag^+ + 4OH^- - - - > 4Ag^o + 4H_2O + N_2$$

The mean size of the silver particles so formed is of order 30nm and the transfer of the silver is achieved by simply immersing the polyester

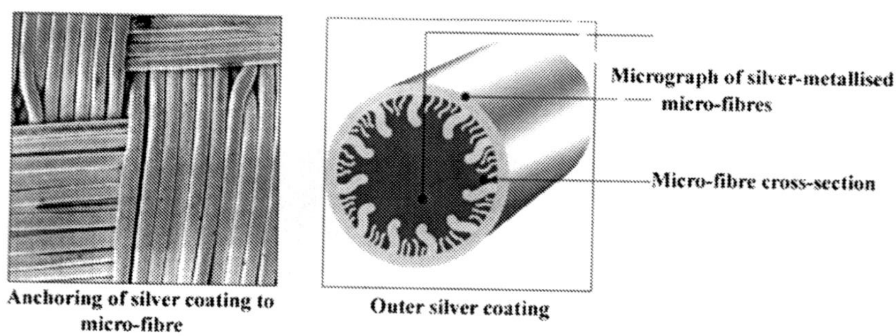

Figure 3.12 Cross-section of a silver-containing fibre and a silver-coated fabric ("Padycare").

fleece in the solution above. The bactericidal action of this material was tested using a Gram-negative bacterium (*Klebsiella pneumoniae*) where a 99.9% destruction of the cells was found (75).

This type of material has also been proposed for use in air filters to inhibit mould and bacteria growth (76). In this case, the proposal was based on use of silver nitrate, sulphate and chloride, compounded with various plastics at a content of 0.1 to 0.5wt%. These were then melt-extruded to form a monofilament fabric. For antibacterial action, a 1wt% silver zeolite in a polypropylene matrix was used, while for fungicidal use, 0.5wt% silver with 10, 10′-oxybisphenoxy was woven for use as an air filter. These fabrics were exposed to *Aspergillus Niger* and *Coliform Bacillus* and the cell count monitored over 72 hours. A significant decay in cell count was recorded (76).

A fairly recent product from Messrs Waveprotect GmbH, a subsidiary of Messrs FEG Textiltechnik GmbH (Aachen) is a functional cloth where cotton fibres are interwoven with silver-coated copper thread (94).

The proteinaceous fibre (Bombyx mori-Silk) is a naturally-occurring amphoteric material with ionisable side-chain functional groupings consisting of various amino-acid radicals. The degree of dissociation of these is a function of pH. These properties allow the silk to adsorb metal ions or other charged species and to bind them (77–83). Thus as soon as the silk is immersed in metal ion-containing solution, it adsorbs these. The rate of this process and its extent, are a function of the metal ion and its valency (or charge), the pH of solution, and time of immersion and solution temperature.

In principle, the protein-metal ion interaction is reversible, although in some conditions (such as low pH), the metal ions are in strong competition with hydrated protons since both species can adsorb or desorb at the same sites. This interaction between metal ions and protein-based fibres, such as silk or wool, has been harnessed in various applications, such as increasing the flame-resistance of wool, or in removal of metal ions from industrial effluents (84).

Loading of silk fibres with silver, to create an anti-bacterial silk-based fabric has been successfully accomplished (82, 85). An extensive study was made of the adsorption and binding of Ag^+ and Cu^{2+} ions to untreated silk fibres as well as those treated with tannic acid and EDTA, as a function of solution pH (85) and the anti-microbial action of the material was exhaustively tested. The silk fibres, both treated and untreated were immersed in an aqueous solution containing 0.5mM silver nitrate with added KNO_3 at 25°C for 30 hrs. pH was adjusted with ammonia or acetic acid. Figure 3.13(a) shows the weight gain of

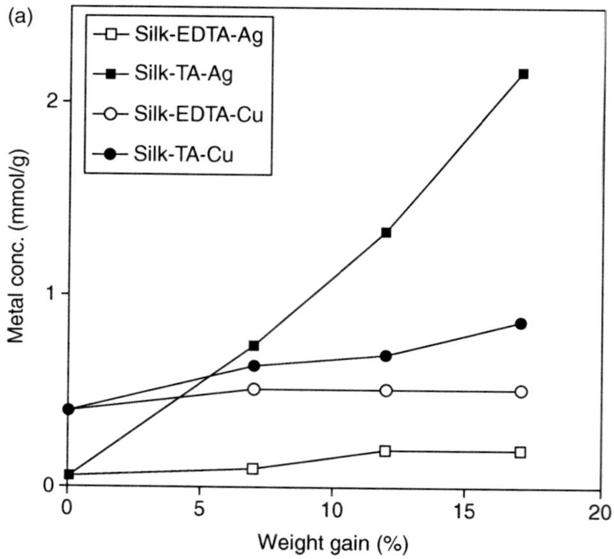

Figure 3.13(a) Uptake of Ag^+ and Cu^{++} ions on treated (tannic acid, EDTA) or untreated silk fibres.

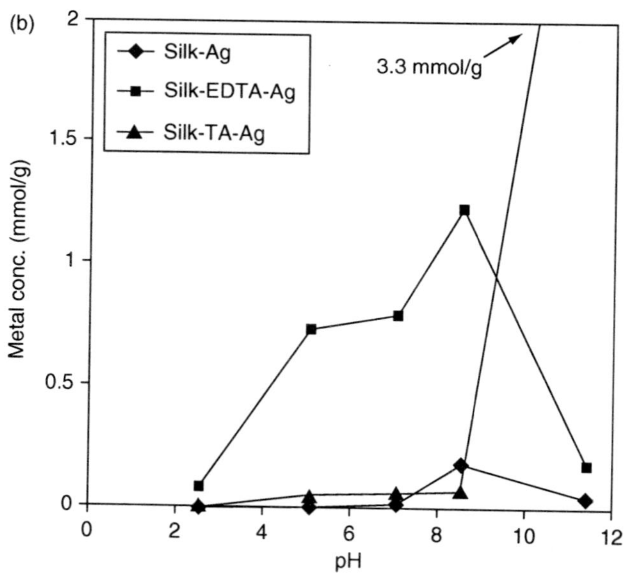

Figure 3.13(b) Adsorption of Ag^+ ions on treated (as Fig. 28a) and untreated silk fibres.

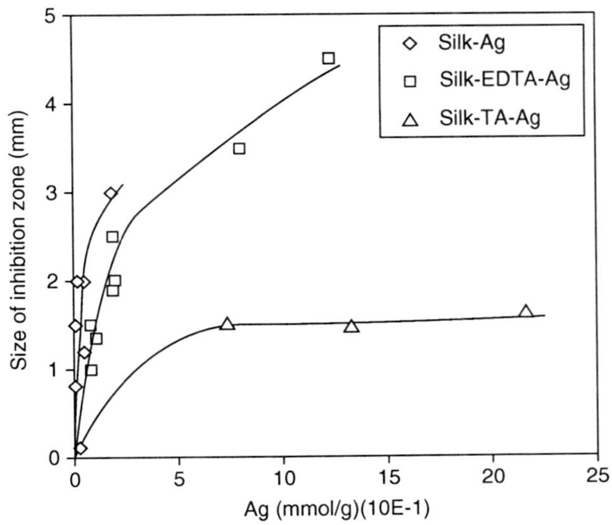

Figure 3.13(c) Size of the cell-free zone as function of Ag^+ ion release in an agar diffusion test.

the fibres after various pretreatments, for copper and silver containing solutions. Figure 3.13(b) shows the effect of pH on silver uptake, while Figure 3.13(c) shows that both pretreated and un-pretreated silk fibres proved effective in their bactericidal action, even at low silver ion loadings. The data report the size of zone in an agar diffusion test which was cell-free.

An interesting result found with silk fibres which had been EDTA-treated showed the greatest silver ion uptake in acid or neutral solutions (Fig 3.13a) whereas tannic acid treated samples absorbed the largest amounts in alkaline media. This is significant in that the silk fibres are not damaged in neutral pH solutions. It also shows how silver ion uptake can be controlled by adjusting solution pH. Another source of silver ions for anti-microbial applications is the chemical precipitation of silver oxy-salts such as [Ag_7NO_{11}] [Ag_2SO_4] or Ag_7O_{11} either as a powder or formed onto fabric (130). Silver oxy-salts are stable at room temperature, but decompose on warming. In one project, a woven fabric of high-density polyethylene (HDPE) was coated with oxy-silver salts. For this, the fabric was initially immersed in a 1% silver nitrate solution for 10 min, making sure that complete wetting was achieved. This was followed by the slow addition, over 15 minutes, of concentrated $(NH_4)_2S_2O_8$ (or its potassium analogue) with continuous agitation. At

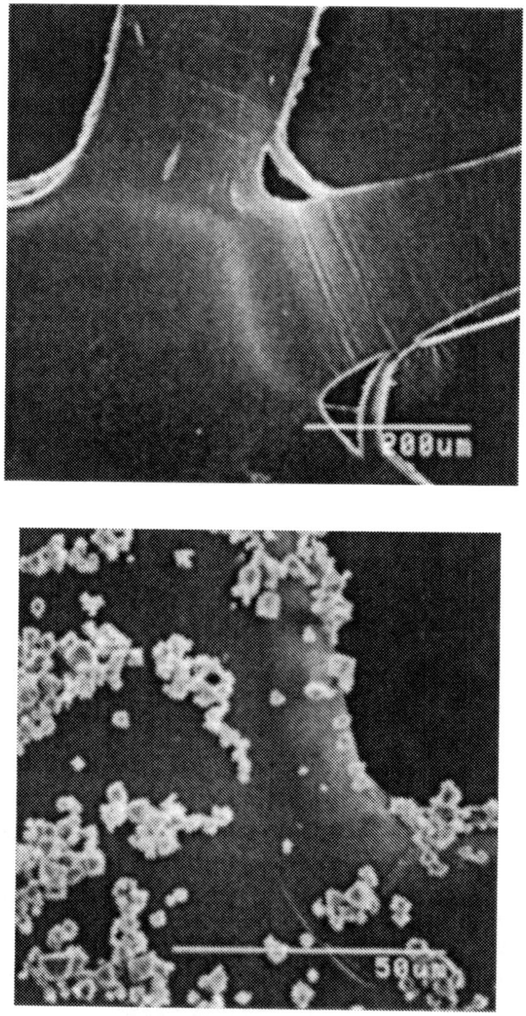

Figure 3.14 SEM images of (top) HD polythene fabric prior to treatment. (below) the grey-black powder deposited on the HDPE fabric (x 1000) (130).

the end of this, the fabric was coated with dark grey crystals (Fig. 3.14 a, b). Finally, the fabric was rinsed with water and dried.

The bioactivity of this coated fabric was tested against various species such as the bacteria *Pseudomonas aeruginosa* and *Staphylococcus aureus* as well as fungi (*Candida albicans*). Standard agar diffusion tests showed that both silver oxy-salt coatings ($Ag_7NO_{11}.3AgSO_4$) and Ag_7O_{11} were effective against these three types of micro-organism (130).

4

Conclusions and Summary

That metal ions can be used to combat the formation and growth of micro-organisms including bacteria, viruses, algae and moulds, is beyond question, as experience and tradition, thousands of years old, in all civilisations, testifies. Since the time of von Nägeli, if not before, this oligodynamic action is underpinned by scientific research. Various studies have demonstrated that it is not the metal itself, which has the microcidal action, but the metal ions formed from the metal, as a result of corrosion or anodic dissolution. It is true that some of these findings appear conflicting, notably where attempts have been made to rank a number of different metal ions in order of their microcidal activity. It seems quite possible that differences in experimental conditions could be responsible for this, and the scientific principles which are normally in evident in corrosion studies, are in many cases absent in studies reported in this book. Until such time as these studies are repeated under more controlled conditions, it is clear that the greatest confidence should be placed in work where micro-organisms were exposed to metal ions in solution, as opposed to studies where such micro-organisms were placed in contact, directly or indirectly, with metals, from which the rate of ion release depends in a complex way, on prevailing conditions. Though a number of metal ions have been shown to have oligodynamic activity, when other factors such as toxicity are taken into account, silver, followed by copper, appear to be the only main contenders.

Metal ions appear to interact in various ways with micro-organisms, for example inhibiting their biosynthesis of proteins. Such is the diversity of micro-organisms, in terms of their structure and the biological pathways by which they survive and propagate, that it would be surprising of all such micro-organisms reacted in the same way to the presence of metal ions. In particular, they vary greatly in the ease with which metal ions can penetrate. As a rule of thumb, Gram-negative species are more susceptible to metal ions than Gram-positive varieties. Resistance to the action of silver ions has been encountered in a few cases. The cause of this appears to be, partly if not entirely, formation of proteins within the cell capable of complexing the silver ions and thus "removing them from circulation". However physical barriers to ingress of metal ions are also a factor to be taken into account, such as formation of slime films surrounding the micro-organism. Such mucous layers can lessen the microcidal action of the metal ions, or indeed totally defeat them.

Notable, both in economic terms and in the wider metabolic picture, is the extremely low concentration at which silver ions can exhibit a microcidal action, of order 0.1μg/l. This should be seen in the context of typical dietary intake of silver, which ranges from 1 to 88μg/day, of which 90% is immediately excreted. The toxicity of silver has been considered in a previous chapter, with external symptoms of argyrosis becoming evident long before any irreversible medical damage takes place. In the writer's opinion, one source of danger one could reasonably anticipate would be ingestion of significant quantities of so-called "colloidal silver", sold on the internet as a panacea without restriction, and directly or implicitly claimed as a cure for anything from cancer to AIDS. There appear to be no reports of silver as a carcinogen or a mutagen.

In normal circumstances, naturally-occurring water can be rendered potable by silver ion concentrations of 0.08 to 0.1mg/l and a treatment time of a few hours. However, as has been described previously, this assumes the absence of other species in the water (which might not themselves be harmful) capable of scavenging the silver ions. Chloride and sulphide ions are inorganic species which could act in this way and in seawater, for example, silver ion concentrations would normally be very low. However this can be circumvented by using a complex salt, as is done by the commercially-available Micropur® tablets based on a silver chloride-sodium chloride complex. Almost any metal (such as steel) can remove silver ions by cementation (q.v), while ferrous ions in solution would be oxidised to ferric, with simultaneous reduction of Ag^+ to Ag^o. . The unfortunate failure of silver-based tablets during the Falklands War (q.v) was cited, where the silver ions were overwhelmed by high concentrations of waterborne animal-derived faecal matter.

The legislative position regarding permitted silver ion concentration in potable water has been considered (Chapter 1) and the present very low EU permitted levels appear to have no logical scientific or medical basis, not least when set against the fact that naturally-occurring silver levels in certain large US cities, are far higher, with no adverse effects ever having been reported. Similar anomalies appear to exist in terms of permitted concentrations of silver in effluents. German legislation demands a limit not exceeding 0.08gm/l of silver ions in effluent, notwithstanding the finding by the Swiss EAWAG that 0.5mg/l Ag^+, and quite possibly more, cause no harm to the bacterial flora used in municipal effluent treatment plants.

4.1 PROSPECTS FOR THE FUTURE

What might the future hold for this branch of science and technology? In the field of water treatment, it is clear that the legislation regarding the use of silver is confused and, in the eyes of many, devoid of logic. However the European Union, which largely controls such legislation in most of Europe, is known for the snail-like pace at which changes are introduced, and few are optimistic that a more logical framework will be introduced any time soon.

On the scientific and technical fronts, in water-treatment, economic pressures will drive two trends. Firstly to ensure that no more silver is dosed into a water supply than conditions demand. Secondly to explore further the extent to which more expensive silver can be supplemented with lower-price copper. As part of the first of these, there will be a continuing search for user-friendly and low-cost analytical methods for copper and silver in solution. Such analyses can reveal both wasteful over-dosing, and also conditions where, for some unexpected reason, metal ions are being removed from a system and thus no longer available for their intended purpose. The concepts of COD (Chemical Oxygen Demand) and BOD (Biological Oxygen Demand) are long-established and linking these, especially the latter, to detect conditions where silver ion disinfection might fail because of too great a burden of general organic matter in a water source, might prove worthwhile.

It is clear from what was written in the relevant sections, that "driven" water-treatment systems, where an externally-applied current is used to control the amount of metal ions passing into solution, must be more accurate and reliable than those where the dissolution process is based on relatively unpredictable conditions such as temperature, pH etc. Yet is an external current source necessary? In thermodynamic terms, the answer is no. The overall process by which a metal ionises and dissolves, does not, in theory, require input of external energy. The process should be seen in terms of battery or fuel-cell technology, where a silver or copper anode is partnered with a cathode. This might be an air-breathing fuel-cell cathode, where oxygen is reduced.. By electronically controlling the resistance of a circuit incorporating these two elements, the rate of metal ion release could be controlled.

Turning to use of surfaces with bactericidal properties, the technology would seem offer scope for further development. At present, the main approach appears to be based on use of a single metal (such as copper door handles) and even silver-loaded stainless steel essentially falls under this heading. As has been noted, the rate at which such metals release their ions must depend on ambient conditions, notably humidity. Whether incorporation of a hygroscopic species could lessen such a dependency remains to be seen. Another approach would be to create a bimetallic surface – in effect a battery for which the necessary electrolytic action would be provided by the moisture present on the human hand that touched it. Dr Giovanni Maggiore (176), discussed present and future applications of nano-silver particles. Apart from applications already mentioned above, he noted equipment for the cheese-making industry, packaging materials, membranes for water treatment, telephone handsets and anti-fouling for boats and ships. A US company, AgION Technologies, of Wakefield, Mass. have developed the use of a silver ion-loaded zeolite for a range of applications, most recently for incorporation in the coating of cheese rinds, to prevent unsightly mould growth.

Finally we come to the use of bioactive textiles. It is true of virtually any branch of technology, that when military R& D funds have been used, a significant advance will be made. Both for soldiers in the field, and also for astronauts, the benefits of clothing with bioactive properties are clear, and it seems highly probable that civilians will benefit from "spin-off" of such R&D.

Last but not least, one notes a trend in pharmaceutical research where many feel that – at least for the present – the law of diminishing returns has kicked in, where development of new drugs is concerned. Huge R&D expenditure seems in so many cases to bring such meagre returns. Perhaps as a reaction to such perceptions, there has been a heightened interest in "folk medicine" and "folk remedies". In many cases, rigorous trials confirm that these are indeed soundly based and not simply placebos. The active ingredient(s) can then be isolated and provide an impetus for new lines of research. Oligodynamically-active metals attract a similar interest. Their action is indisputably real, as millennium-long records show. Yet in circumstances where the costs of medicines and medical care threatens to overwhelm national budgets, these are not intrinsically expensive. The challenge to scientists and technologists in the field is to better understand the mechanisms at work and to optimise the way in which these are utilised. Oligodynamic metals have an ancient past, but an equally bright future.

5

The CD-ROM

The CD-ROM includes an abstract database with over 500 abstracts, also the full text of some key publications. The functionality of this CD allows readers to click straight through to the internet in some cases, or to copy and paste URL's in other cases, thereby accessing a vast range of information. The database itself comes with a powerful search engine, with full Boolean operators. Most of the entries are patents and readers are reminded that the full text of these is available free of charge on the internet for some 99% of all patents, granted or applied for. They include water treatment, both in plumbed-in installations and as portable water containers; air purification and de-odorising; bactericidal surfaces of all kinds, bactericidal paints, household applications including pet toilet facilities. Medical, surgical and dental devices are well represented (use those words as search terms) both as used by the professions and for personal hygiene. Clothing, footwear, fabrics and textiles form the subject of numerous patents, and food and drink-based applications are likewise included. Patents and papers going back to the 1930's will be found, right through to 2007. Technical details relating to the use of the CD-ROM are found in a separate booklet.

References

(Note: Some titles of papers written in German have been translated into English). The CD-ROM database should be considered as a searchable supplement.

1a. Edelmetallschichten, by H. Kaiser, – publ. Eugen E. Leuze Verlag, Bad Saulgau, Germany, 2002, Vol. 31.

1b. Little, M. – Products Finishing, Feb. (2004) 56.

2. Frankfurter Allgem. Z, 15.11.2003.
3. Kawasaki Steel Technical Report No. 46, June 2002, pp. 37 et seq.
4. Facility + Management/Management + Krankenhaus 05/2003, S. 21.
5. Schierholz, J.M. Rump, A.F.E. Pulverer, G. Anaesthesiol., Intensivmed. Notfallmed. 1997, 32, p. 298.
6. Doerr, R. Biochem. Z. 107 (1920) 207: 131 (1922) 351.
7. Acel, D. Z. f. Biochem. 112 (1920) 23.
8. Gibbard, J J. Am. Public Health, 27 (1937) 122.
9. Djokic, S.S. Burell, R.E. – J. Electrochem. Soc. 145 (1998) 1426.
10. Dowling, D.P. Betts, A.J. Pope, C. McConnell, M.L. Eloy, R. Armaud, M.N. – Surface and Coating Technology 163 – 164 (2003) 637–640.
11. Schierholz, J.M. Lucas, L.J. Rump, A. Pulverer, J. – Hospital Infection 40 (1998) 257.
12. Becker-Willinger, C. Csőgőr, Z. Gerwann, J. Schmidt, H. Steingrőver, K. Presentation, – European Coatings Conference, Smart Coatings II, 2003, Berlin.
13. Fromherz, H. Angew. Chem. 50 (1937) 679.
14. Thurman, R.B. Gerba. Ch. P. – Crit. Rev. Environmental Control 18 (1989) 195.
15. Just. J. Sznidis, A. J. Am. Water Works 28 (1934) 451–492.
16. Peterwig, H.G. – Pharmacol. Ther. 1 (1976) 127.
17. Williams, R.L. Grasshoff, C.J. Williams, D.F. – Crit. Rev. Biocompatibil 5 (1989) 221.
18. Rahm, R.O. Landry, R.C. – Photochem. Photobiol. 18 (1973) 29.
19. Luckens, R.J. – Chemistry of Fungicidal Action, Springer-Verlag, New York, 1971 – Structure activity relationship, pp. 76–90.
20. Martin, H. – Fungicides, Vol. 11, Chemistry and Physiology Academic Press Inc. New York, p. 101–107, 1969.
21. Hess & Reiter, Med. Klink 16 (1926) 982 et seq.
22. Angioletto, E. Riella, H.G. Cadore, J.C. Samania, Jr. A. – Quali 2002, VII Congreso Mundial De La Calidad del Azulejo y del Pavimento Ceramico.

23. An overview of the biological action of dental alloy components: Heraeus Kulzer GmbH, Dr. M. Schuster, p. 6, 66003968 (in German).
24. PTAheute (Zeitschrift der Deutschen Apothekerzeitung, DAZ); Nr. 12 Dec. 1996, S. 12–30.
25. Deutsche Apotheker Zeitung; Vol. 139, No. 46 (1999) p. 48.
26. Ibid Vol. 138, No. 7 (1998) p. 41.
27. Nakamura, S. Ookubo, N. Miyakusu, K. Hasegawa, M. Munegue, Y. – Nisshin Steel Technical Reports 76 (1997) 48.
28. Nishimura, T. – "The Basic Knowledge of Antibacterial Activity" (1999) (Techno System).
29. Yokota, Y. Tochihara, M. Ohtam, M. – Kawasaki Steel, Technical Report No. 46, June 2002.
30. Massalski, T.B. – Binary Alloy Phase Diagrams, ASM International (1988) 35.
31. Japanese Standards Association – Antimicrobial Products – Test for Antimicrobial Activity and Efficacy, JISZ 2801, (2000).
32. Christensen, G.D. – Hosp. Practice 1993, 30, pp. 27–36.
33. Costerton, J.W. Cheng, K.J. Geesey, G.G. Ladd, T.I. Nickel, J.C. Pasgupta, M. Marrie, T.J. – Am. Rev. Microbiol. (1987) 41, S. 435–464.
34. Schierholz, J.M. Jansen, B. Jaenicke, L. Pulverer, G. – Biomaterials (1994) 15, S. 996–1000.
35. Sheth, N.K. Franson, T.R. Sohnle, P.G. – Lancet (1985) 2, pp. 1266–1268.
36. Management + Krankenhaus 11/2003, p. 28.
37. Kahl, K. Grosse-Siestrup, C. de Veer, I. Becker, H. Ruden, H. Hahn, H. et al. Int. Conf. On Prevention of Infection, Nice FR. (1992).
38. DE 4115390Al.
39. Raad, I. Hachem, R. Zermeno, A. Dumo, M. Bodey, G.P. – Biomaterials 17 (1996) 1055–1059.
40. Scott, O.T. Johnson, M.S. Shah, H. Kraus, M.A. McKusky, M.A. Ambrosius, W.T. Harris, V.J. Snidow, J.J. – Radiology May 1998, pp. 491–496.
41. Guggenbichler, J.P. Boswald, M. Lugauer, S. Krall, T. – Infection 27 (1999) 16.
42. Mahan, J. Seligson, D. Henry, S. Henry, P. Dobbins, J. – Factors in Pin Tract infections. Orthopedics. 1991, 14, 408 8.
43. Wassal, M. Santin, Isalberti, C. Cannas, M. Denyer, S. – J. Biomed. Mater. Res. 1997, 36, pp. 325–30.
44. Colline, C. Goll, G. Seligson, D. Easley, J. – Orthopedics, 1994, 17, pp. 445–8.
45. Masse, A. Bruno, A, Bosetti,M. Biasibetti, A. Cannas, M. Gallinaro, P. – Biomed. Mater. Res. (Appl. Biomater) 2000, 53, 600 4.
46. Bosetti, M. Masse, A. Tobin, E. Cannas, M. – Biomaterials 23 (2002) 887–892.
47. Daunderer, M.: Handbook of Eco-Toxins. Specific Information on Silver III-3. Supp't. 5/93, (in German). Publ. Ecomed-Verlag, Landsberg-Lech 1993.
48. Zapata-Sirvent, R.L. Hansbrough, J.E.; J. Burn Care Rehabil. 14 (1993) 132–140.
49. Boosalis, G.M. McCall, J.T. Ahrenholz, D.H. Solem, L.D. McCain, C.J. – Surgery 101 (1987) 40–43.

50. Wood, P.N. New Antimicrobial silver formulations, 2nd European Precious Metals Conference Lisbon, 10–12 May 1995.
51. D.E. Pat. No. 10054248, 2.11.2000.
52. D.E. Pat. No. 10128625, 13.6.2001.
53. D.E. Pat. No. 19645043, 19645043.
54. Svanberg, M. Krasse, Örnerfeldt, H.O. – Caries Res. (1990), 24, 133–36.
55. Yoshida, K. Tanagawa, M. Atsuta, M.; J. Biomedical Materials Research (1999), 516–522.
56. Wang, W. Efrima, S. Regev, O.; J. Biomedical Materials Research (1999) 516–522.
57. Lee, C.P. Meisel, D. – J. Phys. Chem. 1982, 86, 3391.
58. Michaelis, A.M. Nirmal, M. Brus, L.E.; J. Am. Chem. Soc. 1999, 121, 9932.
59. Oku, T. Shigeru, K; Yazawa, T; Ijima, T. Kumakura, N. – US Pat. 5807641.
60. Sugiura, K. Maekawa, S. Inoue, H. – Oomura, T. Kato, H. Kourai, H. – US Pat. 5296238, 1994.
61. Atsumi, K; Saito, T. Komori, M. – US Pat. 5151122, 1992.
62. Prat, A.S. Smith, P.R. – US Pat. 4849223, 1989.
63. Ahmed, A.A. Abdullah, E W. – Phys. Chem.. Glasses, 38 (1997) pp. 42–50.
64. Navarro, J.M.F. – El Vidrom Consejo Superior de Investigationes Cientificas, 2nd Ed. Madrid, 1991.
65. Berger, T.J. Spardaro, J.A. Bierman, R. Chapin, S.E. Becker, R.O. – Antimicrobial Agents and Chemotherapy, Vol. 10, No. 5, 1976, pp. 856–860.
66. Borsella, E. Gonella, F Mazzoldi, P. Quaranta, A. Battaglin, G. Pollini, R. – Chemical Physics Letters 284 (1998) pp. 429–434.
67. Sharaf, N A. Sr. Condrate, R.A. Ahmed, A.A. – Materials Letters 11 (3/4), 1991.
68. Bellantone, M. Hench. L.L. – Key Engineering Materials Vols. 192–195 (2001) pp. 617–620.
69. Cao, W. Hench, L.L. Ceram. Internat. 22 (1996) pp. 493 et seq.
70. Kawashita, M; Toda, S. Kokubo, T. Masuda, N. – Key Engineering Materials Vols. 240–242 (2003) pp. 229–232.
71. Doremus, R. H. – Glass Science (Wiley-Interscience), New York (1994) p. 20.
72. www.uniklinkum-freiburg.de/index.xml; Schuster, A.
73. "Reduction of Staphylococcus-aureus infection using Padycare® silverised Textiles on Patients with Atopical Eczema" Prof. Dietrich Abock TU Munich. Jan. 2003, Prof. Dr. Guggenbichler, Munich (in German).
74. Brochure from Messrs Tex-A-Med GmbH on Padycare®.
75. Lee, H.J. Jeong, S.H. Yi, S. – The Fiber Society Fall 2001 Annual Technical Conference 30.10 – 1.11.2001, Lake Tahoe.
76. US. Pat. 6165243, 26 Dec. 2000.
77. Shimizu, F. Aida, G. – Sen-I Gakkaishi, 1970, 26, 316.
78. Sugiura, M. Sen-I Gakkaishi, 1973, 29, T-220.
79. Shimizu, M. Jyoko, K. Sakaguchi, I. – Sen-I Gakkaishi 1976, 32, T-529.
80. Shimizu, M. Sakaguchi, I. Sen-I Gakkaishi, 1980, 36, T-127.
81. Nakamura, Y. Kurose, A. Shimiu, F. Shirai, H. Hojo, N. – Sen-I Gakkaishi, 1985, 41, T-353.
82. Chen. W. Koyama, T; Hanabusa, K. Shirai, H. – Sen-I Gakkaishi 1995, 51, T-176.

83. Shimizu, Y. Takagishi, T. J. Seric Sci. Jpn 1997, 66, 1.
84. Masri, M.S. Reuter, F.W. Friedmann, M.; Text. Res. J. 1974, 44, 298.
85. Arai, T. Freddi, G. Colonna, G.M. Scotti, E. Boschi, A. Murakami, R. Tsukada. M. – J. Applied Polymer Science (2001) pp. 297–303.
86. Kleine Enzyklopädie-Technik, 1958, Verlag Enzyklopädie Leipzig, S. 611.
87. www.inflitec.de.
88. Biomedicine – Synergies create Inventions; phi (Produktionstechnik. Hanover informiert); 4 Aug-Oct 2003, p. 4.
89. Von Nägeli, K; "On Oligodynamic Phenomena in Living Cells" Neue Denkschr. Allgem. Schweiz Ges. Naturwiss; Vol. 33, pp. 174–182, (1893).
90. Decker, S.M.: Dissertation Rheinisch Friedrich-Wilhelm Univ. Bonn; (1996).
91. German Patent DE 4324994Cl: also PCT/DE2007/000195 "Oligodynamisch wirksame Vorrichtung und Verfahren zu ihre Herstellung.
92. European Patent EP 0636375Bl.
93. Frankfurter Allgem. Zeitung No. 47, 25.2.2003, Supp't. T6.
94. Waveprotect GmbH, NRW, D-52382 Niederzier, Germany.
95. Metallic Biomaterials – Biehl, V. Breme, I. Mat.-wiss u. Werkstofftechnik Vol. 32, pp. 137–141 (2001).
96. Van Oeveren, W. Schoen, P. Maijers, C.A. – Progress in Biomedical Research, May 2000, pp. 211–214.
97. "Overview of the biolgical effects of the components of dental alloys" (in German); http://www. Biopartner.de/legierunbio.
98. Degussa Precious Metals Handbook; p. 606, Huthig-Verlag Heidelberg, 1995 (in German).
99. Addicks, L. – Silver in Industry, Reinhold Publishing Corp., New York, 1940,pp. 401–450, 584–597.
100. Oberzill, W. "Oligodynamic action of silver and its use in disinfectation of drinking and mineral water"; publ Scientia Pharmaceutica, Vienna (1956), pp. 171–193.
101. Water Quality and Treatment, 2nd ed. J. Am. Water Works Ass. New York, 1951.
102. Copper & Cuprous Materials; A Commentary on Public Health Issues (in German). German Copper Institute. Order No. i. 28, 09/97.
103. Primer on Potable Water; Katadyn (Switzerland) Produkt. AG.
104. Rems- Zeitung; (Schwäbisch-Gmünd) 8.7.1999, ibid. 19.3.1999.
105. Management & Krankenhaus 06/2004, p. 27.
106. Management & Krankenhaus 11/2004, pp. 33–34.
107. Management & Krankenhaus 06/2004, p. 1.
108. Oberzill, W. Experimental Basis of Silver-Dosing Potable Water (in German) Arch. Hyg. Bakt. 1957, pp. 217–234.
109. der-brunnen.de/infos ũber Wasser.
110. Galvanotechnik 81 (1990) p. 3526.
111. Hurley, T.L. et al.. Potable Water Bactericide Agent Development Chemitric. Inc. Rosemont, Ill, Final Report on NASA Report No CR-115595, (1972) Doc. ID 19720020458. Acc'n No: 72N28108.
112. Requirements for Ion-Exchangers; (in German); Swiss Ass'n of Gas & Water Engineers, Zũeich 30.9.1977.
113. Degussa-Mitteilung 87/75, Sept. 1975.

157. Wysor, M.S. – Orally-administered Silver Sulfadiazin, Chemotherapy and Toxicology in CF-1 Mice, Plasmodium berghei (Malaria) and Pseudomonas aeruginosa. Chemotherapy 21 (1975), pp. 302–310.
158. Wysor, M.S. Zollinhofer, R.E. – Antibacterial Properties of Silver Chelates of Uracil and Uracil Derivatives in vitro. Chemotherapy 17 (1972), 188–199.
159. Wiedemann, B. Antibiotics & Nosocomial Infections (in German) (in) Bacterial Nosocomial Infections. Publ. C Krasemann & W Margert, Verlag Walter de Gryter, Berlin, New York, 1986, pp. 61–68.
160. U.L: personal communication from Forschungsinst. F. Edelmetalle und Metallchemie, Schwäbisch Gműnd.
161. Drescher, H; Development of a fine-pored, high-strength bioceramic for antibacterial use in infusion systems (in German) Diploma Thesis. TU Clausthal 1995.
162. Globulovich, V.N. Rabotnova, I.L. – Kinetics of growth inhibition by silver ions, Microbiology 43 (1974), pp. 948–950.
163. Hendry, A.T. Stewart, I.O. – Silver resistant Enterobacteriacea from hospital patients, Can. J. Microbiol. 25 (1979) 915–921.
164. Chambers, C.W. Proctor, Kabler, P.W. – Bactericidal effects of low concentrations of silver, J.A. Water Works Assoc. 54 (1962) pp. 208–214.
165. Slawson, R.M. Lee, H. Trevors, J.T. – Bacterial interactions with silver. BioMetals 3 (1990) pp. 151–154.
166. Electric Materials, SILVERON AG-100, Rohm + Haas.
167. Vashkov, V.I. – Appliances and Methods of Sterilization used in Medicine, Moscow, (1973). (in) Rev. Physicochemical Properties & Biological Characteristics of some Metals used in Medical Instruments". Ed. F.G Studenikina, V.N.Kosova & V.P.Denisova (in Russ.).
168. Source; German Office of Statistics; Sepsis Network http://destatis.de.
169. Planet Earth; Noble Metals; Time-Life-Verlag (Amsterdam) by. St. John, J. (1971) ISBN 90-6182-496-6; p. 29.
170. Berger, T J; Spadaro, J A; Chapin, S.E. Becker, R.D. – Antimicrob. Agents Chemother, 1976, 9, pp. 357–358.
171. Kaltofen, R; Opitz, R; Schuhmann, K. Ziemann, J. – Book of Chemical Tables (in German), VEB Deutscher Verlag for Basic Chemicals (1986).
172. Grosheger, G. Hardes, J. Ahrens, H. Streitburger, A. Buerger, H. Erren, M. Gunsel, A. Kemper, F.H. Winkelmann, W. Eiff. Ch.; Biomaterials 25 (2004) pp. 5547–5556.
173. Wilks, S A; Michels, H T; Keevil, C W; Int'l. J. Food Microbiol. 111 (2006) 93–98.
174. John Hayes, TarnPure Ltd. Private communication.
175. A search of Google and the British Library on-line catalogue at www.bl.uk for the terms "sustained release" or "controlled release" yields around 1000 books, conference proceedings etc.
176. Maggiore, G; "Soluzioni innovative per l' igiene da nanoparticelle d'argento". Pitture e Vernici, European Coatings 16/2006 (www.gbp.it). Also as a shorter version in besser lackieren (7) 20th April 2007 as "Antimikrobiell und beständig: Nanopartikel aus Silber".
177. Haldar, J; An, D; Alvarez de Cienfuegos, L; Chen, J; and Klibanov, A M; "Polymeric coatings that inactivate both influenza virus and

pathogenic bacteria". Proceedings of the National Academy of Sciences (2006); Vol. 103, Issue 47, November 21,pp 17667-71. Full text avail. at http://www.pubmedcentral.nih.gov/picrender.fcgi?artid=1693804&blobtype=pdf.

178. Park, D; Wang, J; and Klibanov, A M; "One-step,painting-like coating procedures to make surfaces highly and permanently bactericidal" Biotechnology Progress (2006) Vol. 22, (2), pp 584–9. Abstr. Avail. at PubMed.
179. Professor C W Keevil, private communication.
180. Pourbaix, Marcel; "Atlas d'equilibres electrochimiques" publ. Gauthier-Villars, Paris, 1963. Re-published several times thereafter, in both French & English language versions.
181. Robine, E: Boulange-Petermann, L; et al; "Assessing bactericidal properties of materials: the case of metallic surfaces in contact with air" J. Microbiol. Methods 49 (2002) (3) 225–234.
182. Herbert Käszmann, private communication.
183. Rudnev, V S, Tyrina, L M; Nikitin, V M, et al; Protection of Metals 39(4) (2003), 334–337.
184. Brain, Steven, PhD thesis. University of the South Bank London 1995/6 "Electrochemical Techniques for Study of Biofilms".
185. Noyce, J O; Michels, H; Keevil, C W; Appl. Environm,. Microbiol (2007) (4), 2748–2750.

Index

Printed in the United Kingdom
by Lightning Source UK Ltd.
125053UK00003B/109-600/A

9 780904 477306